THE BIG BOOK OF

SCIENCE

Brimming with creative inspiration, how-to projects, and useful information to enrich your everyday life, Quarto Knows is a favorite destination for those pursuing their interests and passions. Visit our site and dig deeper with our books into your area of interest: Quarto Creates, Quarto Cooks, Quarto Homes, Quarto Lives, Quarto Drives, Quarto Explores, Quarto Gifts, or Quarto Kids.

This edition published in 2018 by Chartwell Books,
an imprint of The Quarto Group
142 West 36th Street, 4th Floor
New York, NY 10018 USA
T (212) 779-4972 F (212) 779-6058
www.QuartoKnows.com

Conceived, designed and produced by Quid Publishing,
an imprint of The Quarto Group,
The Old Brewery,
6 Blundell Street,
London N7 9BH
United Kingdom

Chartwell Books titles are also available at discount for retail, wholesale, promotional, and bulk purchase. For details, contact the Special Sales Manager by email at specialsales@quarto.com or by mail at The Quarto Group, Attn: Special Sales Manager, 401 Second Avenue North, Suite 310, Minneapolis, MN 55401, USA.

10 9 8 7 6 5 4 3 2 1

ISBN: 978-0-7858-3599-8

Page design by Lindsey Johns
Printed in China

For Barnaby—voila

THE BIG BOOK OF
SCIENCE

Facts, figures, and theories to blow your mind

CHARTWELL BOOKS

JOEL LEVY

Contents

01

Physics

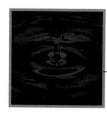

02

Chemistry

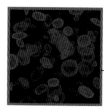

03

Biology

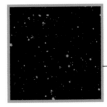

04

Astronomy

05

Earth Science

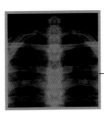

06

The Human Body

07

Technology

Introduction

To discover how the layered structure of Earth is analogous to that of a Scotch egg, turn to page 174.

▶ Did you know Earth is like a Scotch egg, or that if cars were like computers you could get to the Moon and back on a single tank of gas? Are you curious about Maxwell's demon and Descartes' evil genius? Did you know that you are like a Flatlander, with a little bit of Shakespeare and Genghis Khan thrown in? Would you like to know how a spider could catch a jumbo jet, or find out what happened to Schrödinger's poor cat?

This book will tell you all these things and more: how many elephants the Sun burns in a second, and why being a sperm is like swimming in a pool filled with molasses when you can only move your arms a centimeter a minute; why an electron is like a bee in a cathedral, and how you could age twice as fast as your twin or live longer by sitting in a hole; how long it would take to drink the Pacific or get to the top of Mt Everest in an elevator.

This stew of analogies and thought experiments is intended to amuse your intellectual palate, but hopefully its rich gravy of metaphor and simile will also get you to swallow some of the most difficult concepts in science by making them just a little easier to digest.

Anatomy of an analogy

Science can be hard for all sorts of reasons; it may involve difficult mathematics and complex equations, deal with the incomprehensibly vast or the unimaginably minute, use jargon and notation, or appear to defy logic and intuition. For all these reasons and more, science educators and communicators have traditionally turned to analogies, metaphors, similes, and thought experiments to explain and clarify scientific ideas.

Analogies have great power because they tap into the architecture of the human mind. The mind evolved in a complex and fluid environment and has specialized to process often partial and confusing information in a quick and effective manner. It uses rules of thumb and takes shortcuts wherever possible, which is

An analogy is made up of a source or **analog** (an accessible and often familiar concept, situation, or object) and a **target** (the concept that is being explained or compared to). Links between the source and the target are known as mappings. Positive mappings are shared attributes, ways in which source and target are alike; negative mappings are ways in which the target is not like the source—in other words, instances where the analogy becomes misleading.

one reason why people tend to be poor at following strictly logical thought processes. Instead we find it easier to reason by comparing unfamiliar with familiar, falling back on experience, looking for links between things, and seeking out pattern and meaning. Reasoning by analogy takes in all of these processes and more; indeed, analogy is so powerful that it is central not only to the communication of science but also to the process of scientific advancement itself. Analogy is a key element of the mysterious phenomena of scientific inspiration and creativity, and the history of science is filled with examples of breakthroughs achieved via analogous reasoning.

Revolution by analogy

It is only necessary to look at the defining moments of the birth of the Scientific Revolution to see analogy in action. Johannes Kepler elucidated the laws of planetary motion—among the very first laws of science. The achievement sprang from his initial conception of the universe as a giant clockwork mechanism (see pp. 10–11), an analogy that gave him the conviction to challenge established theories of the world and the impetus to discover the mathematical laws of the cosmos. His work—and his analogy—in turn inspired a young Isaac Newton to compare the fall of an apple in his mother's garden to the orbit of the Moon, and to wonder whether some underlying principle meant these processes were more than merely analogous.

There are dozens more examples of analogies inspiring scientific discoveries, many of which are covered in this book. Robert Boyle, for instance, was inspired to develop his theories about gases by imagining gas particles as coiled springs (see pp. 38–9). Christiaan Huygens and many after him analyzed light phenomena by analogy with waves in water. James Clerk Maxwell modeled electric lines of force as water pressure in tubes (see pp. 48–9). August Kekulé was inspired to describe the ringlike structure of the compound benzene by a dream of a snake biting its own tail. Watson and Crick, and thousands of scientists before and since, used a model—a form of

analogy—to help them arrive at the double helix structure of DNA (see pp. 90–1).

Thought experiments are a form of reasoning by analogy or metaphor. They have proved their utility in most fields of science, but rarely as effectively as in physics, where many of Einstein's epochal breakthroughs were achieved by means of thought experiments such as "what would it be like to ride on a beam of light?" or "what forces does a falling man feel?" By considering these hypothetical situations, Einstein arrived at his theories of relativity and redefined our understanding of the universe (see pp. 14–15).

When analogies go bad

Awareness of the limits of analogy is essential because, like any powerful tool, analogy can be misused and abused. Negative mappings can be misleading, and can lead to entrenched misconceptions. A common example is the way in which the analogy between electricity and water (see pp. 48–49) leads to some false beliefs about electricity—for instance, that it might leak out of an unplugged cable. A much more serious example is the way in which a misunderstanding of Darwin's analogy between evolution and a tree (see pp. 100–101) has led to a widespread misconception that evolution is a progressive force, implying a natural hierarchy or "ladder" of life. This mistake underpinned generations of racist pseudoscience with horrible consequences, from the colonial subjugation and genocide of aboriginal races to the bogus eugenics of the Nazis. Faulty analogies can also be used to obscure and block scientific thought, as in the superficially powerful watchmaker analogy employed by the creationist, or intelligent design, movement (see pp. 126–7).

In general, however, analogies play a positive role in science, encouraging wider discussion of otherwise ghettoized concepts and adding a touch of fun to a field often perceived as dry and dusty. After all, science is like a houseplant—it needs to be taken out of its dingy corner and put in the sunlight once in a while if it's to flourish.

Section One

▶ Physics is the study of the fundamental forces of the universe; it deals with energies beyond comprehension, physical and temporal dimensions beyond imagination, concepts of dizzying complexity that run contrary to common sense, and mathematics—lots and lots of very hard mathematics. It is the field that most needs, and most benefits from, analogies, and this section introduces many key physics analogies in quantum theory, relativity, string theory, and other areas.

Physics

Kepler's clockwork cosmos

▶ *"The machinery of the heavens is ... like a clock ... all the variety of motions is from one simple force ... as in the clock all motions are from a simple weight."* Johannes Kepler

According to the premodern conception of the cosmos, the universe was composed of a succession of concentric, perfect crystal spheres, with Earth at the center. This geocentric system owed more to mysticism than to astronomy, assuming some sort of mystical correspondence between the various spheres, governed by mysterious laws and unseen forces. In order to "rescue" the perfect nature of the spheres, complex mathematical contortions were required to match this theory to the motions of the celestial bodies as they were actually observed in the night skies.

In the 16th and 17th centuries Nicolaus Copernicus (1473–1543) and Johannes Kepler (1571–1630) looked at the evidence and suggested a much simpler, more elegant heliocentric system, with the Sun at the center of the cosmos and the planets orbiting around it. What was truly revolutionary about this approach was that it explained planetary motion according to simple, mathematically stated laws, which were universal—in other words, which applied to all celestial bodies. The cosmos had the elegance and beauty of clockwork, a great machine of celestial cogs and gears. What was more, Kepler guessed (but could not prove) that this celestial clock was driven by a single force, just as a pendulum clock is driven by the swinging of a weight. It was left to Newton to show that this force was gravity.

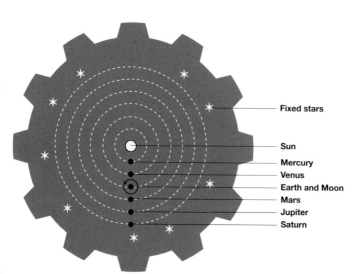

Fixed stars

Sun
Mercury
Venus
Earth and Moon
Mars
Jupiter
Saturn

The Copernican model of our Solar System consisted of eight spheres: six for the known planets orbiting the Sun, one for the Moon orbiting Earth, and one final sphere for the distant stars.

1,000
Number of named stars known in
ancient times ✳

2,000
Number of named stars known
*c.***1600 CE** ✳ ✳

3,000
Number of named stars known in
1712 ✳ ✳ ✳

225,300
Number of named stars known in
1918 ✳ ✳ ✳ ✳

16 million
Number of named stars known in
1983 ✳ ✳ ✳ ✳ ✳

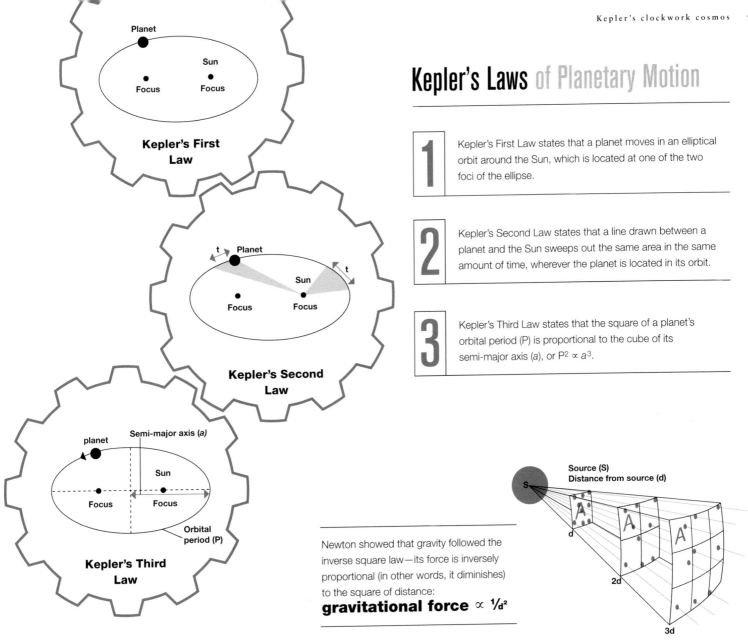

Kepler's First Law

Kepler's Second Law

Kepler's Third Law

Kepler's Laws of Planetary Motion

1 Kepler's First Law states that a planet moves in an elliptical orbit around the Sun, which is located at one of the two foci of the ellipse.

2 Kepler's Second Law states that a line drawn between a planet and the Sun sweeps out the same area in the same amount of time, wherever the planet is located in its orbit.

3 Kepler's Third Law states that the square of a planet's orbital period (P) is proportional to the cube of its semi-major axis (a), or $P^2 \propto a^3$.

Newton showed that gravity followed the inverse square law—its force is inversely proportional (in other words, it diminishes) to the square of distance:

gravitational force $\propto \frac{1}{d^2}$

Source (S)
Distance from source (d)

Kepler was a **mathematician**, not an astronomer. For the astronomical data he needed to develop his laws of planetary motion, he depended on the great Danish astronomer Tycho Brahe. Brahe was a colorful character; he had lost the end of his nose in a duel and wore a fake nose of silver-gold alloy. He also had a pet moose that died after it got drunk on beer and fell down the stairs.

Why you can never unshuffle a pack of cards

▶ Shuffling a freshly unwrapped pack of cards disorders the suits, and you cannot get them back into order simply by shuffling again and again. This is entropy in action.

The second law of thermodynamics states that the entropy of a closed system never decreases. Entropy is a measure of order, or rather of disorder; it is a concept that applies to both information and energy. A fresh pack of cards has relatively low entropy: all the suits are together and in order. Shuffling them once starts to break up this pattern, introducing disorder and increasing the entropy of the system. Repeated shuffling will not put the cards back into order; rather, the more you shuffle, the more mixed up the suits and numbers become, as entropy increases. In fact this analogy is not entirely accurate; if the cards were shuffled for an infinite amount of time the starting configuration would eventually be reached by chance. A more accurate analogy might be with a Rubik's cube that starts off with each side a solid block of color and is then twisted randomly. The order of the colored blocks quickly breaks down and a lifetime of random twisting will not get it back to its original state.

Equivalent to these two analogies is a container with two compartments, one filled with hot gas and the other with cold gas. According to the laws of thermodynamics, heat energy will flow from the hot side to the cold side, until the temperature of the two compartments has equalized; as it does so the entropy of the system increases. It is impossible to reverse this process.

Maxwell's demon

Physicist James Clerk Maxwell proposed a thought experiment (shown in the diagram below) in which an entity—later called "Maxwell's demon"—controls a little door between the compartments, allowing only high-energy particles passage from the left-hand compartment into the right-hand one, and vice versa, decreasing entropy and breaching the second law of thermodynamics, like shuffling a disordered pack of cards back into suit. In fact, the demon constitutes a source of energy and entropy himself, and once he is included in reckonings of the system, the second law is preserved.

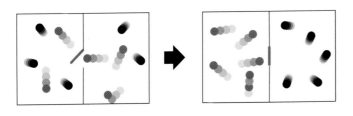

Perpetual motion machines are impossible because they would violate the second law of thermodynamics. The nearest thing to it is a vacuum-sealed magnetically levitated flywheel.

Longest running flywheel:

12 years

Based on the assumption that it will continue to expand, it is predicted that the universe will reach maximum **entropy** in

$10^{100} - 10^{1,000}$ **years**

10^{10}

Current **age** of universe:

x1.38 **years**

Physicist Arthur Eddington pointed out that the chance of the second law of thermodynamics being breached is considerably less than the chances of an army of monkeys accidentally typing "all the books in the British Museum."

If the entire universe were filled with monkeys typing at random until the end of time, the chances of a single **monkey** exactly reproducing Shakespeare's *Hamlet* at the first attempt is

1 in $10^{183,800}$

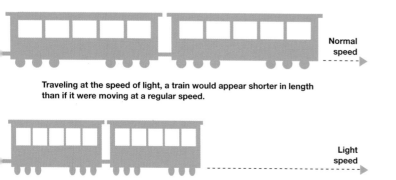

Traveling at the speed of light, a train would appear shorter in length than if it were moving at a regular speed.

Normal speed

Light speed

If you stand on a platform and watch a train go past from left to right, in which two passengers are playing table tennis, the ball will always be traveling left to right, even when the player on the right is returning the ball. To the players, however, the ball appears to be moving back and forth because their frame of reference is different to yours. The speed of the ball is relative to the observer. This is known as Galilean relativity. If the speed of light (known as c) were relative to the observer, light leaving Earth in the direction in which the planet is moving through space would travel faster than light leaving Earth in the other direction.

Short, fast trains in slow motion

▶ Because the speed of light is constant (in a vacuum), for anything traveling at near-light-speed, time dilates and space contracts, so that to an observer standing on a platform, a train passing at near-light-speed would seem shortened, while its passengers would appear to be moving in slow motion.

The Shanghai Maglev train, or Shanghia Transrapid, which runs between Shanghai city and Pudong International Airport, is the world's **fastest** passenger train, with a top speed of

431 kilometers
268 miles
per hour

In 1881 the American physicist Albert Michelson measured precisely these two quantities and found that light moved at the same speed in both directions; in other words, its speed is not relative to the observer—it is an absolute.

In 1905 Albert Einstein realized that if the speed of light was not relative to the observer, then time and space must be relative instead. Imagine that the train going past you is moving at near-light-speed, say at 0.6 *c*. On board, a passenger aims a flashlight at the ceiling; from inside the train the light appears to travel vertically in a straight line. Viewing this from the platform, however, the light seems to you to follow an inverted V-shaped path: it travels further than it does for the passenger, but the speed of the light itself is the same for both of you. The only way to resolve this paradox is to say that, from your point of view, the light beam's journey as viewed by the passenger took less time than it did for you. Her clock is running slower than your clock, from your point of view. Because motion is relative, however, as far as the passenger is concerned it is you who is zipping past at 0.6 *c*, so for her it is your clock that is slower. As time dilates, distance contracts. As a result, to the observer on the platform, the train will appear shortened in the direction of travel. The faster it goes, the shorter it will get and the slower time on board will appear to run.

The **fastest object** ever made by humans is the *Helios 2* solar-mission satellite. It reached a speed of

68.75
kilometers
persecond

Speed of light in a vacuum (*c*):

299,792 km/s

Cosmic ray **muons** are incredibly short-lived, near-light-speed particles created when high-energy cosmic rays collide with the upper atmosphere, which make it all the way to Earth's surface thanks to relativistic time dilation.

Average lifespan of a muon:

0.0000015 **seconds (1.5 microseconds)**

The distance traveled by light in the lifetime of a muon:

457 **meters**

The distance actually traveled by a cosmic ray muon:

~12.5 km **Approx. time dilation factor:**

✕1.5million

is the number of times faster a maglev train would need to go to reach 0.6 *c*.

The elevator and the rocket ship

▶ An astronaut floating in space shines a beam of light through the window of a rocket ship accelerating past. To the passenger in the rocket ship the beam appears curved, yet he believes himself to be in a stationary elevator on Earth.

Gravity causes the ball to accelerate toward the elevator floor.

A dropped ball behaves the same in either scenario.

Firing of the rocket's motor causes the floor to accelerate toward the ball.

The elevator in freefall accelerates at the same rate as the ball.

The two scenarios are equivalent.

The rocket is stationary in space, so the ball experiences no acceleration.

Einstein's principle of equivalence states that a passenger in a windowless, silent rocket ship accelerating through space at a rate equivalent to Earth's gravity (9.8 m/s^2, or 1 g) has no way of knowing that he is not in an elevator on Earth. To put it another way, someone in an elevator that has broken its cables and is plunging down the shaft feels the same force as an astronaut in a rocket ship that is floating stationary in space. In other words, gravity and acceleration are equivalent.

Now imagine that an astronaut floating outside the passing rocket ship shines a laser through a tiny porthole that has helpfully opened up. From the floating astronaut's point of view the laser beam is traveling in a straight line, but the passenger sees something different. Because the rocket ship is moving, the laser does not strike the far wall directly opposite the porthole, but a little below. If the beam were visible he would see it following a curved trajectory. Recall that acceleration is equivalent to gravity, meaning that someone standing in an elevator on Earth's surface would also see the laser beam following a curved trajectory (although because

Zero motion

When the rocket is not moving relative to the astronaut floating outside, the laser beam appears straight to both observers.

Constant velocity

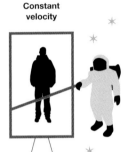

When the rocket is moving at a constant speed, the passenger sees the laser beam slanting down in a straight line.

Acceleration

When the rocket is accelerating, the passenger sees the beam following a curved path, while the astronaut sees a straight beam, relative to which the rocket ship is moving.

gravity on Earth is relatively weak, the curvature would be infinitesimal). Light travels the shortest path between two points, and a curve is only the shortest path on a curved surface. The conclusion is that gravity/acceleration is the curvature of space—to the observer, gravity/acceleration appears to be bending the light. As special relativity shows, time and space are aspects of the same thing, so in fact it is space–time—a four-dimensional construct—that is curved.

Proportion of Earth's gravity

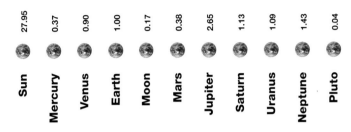

Sun	Mercury	Venus	Earth	Moon	Mars	Jupiter	Saturn	Uranus	Neptune	Pluto
27.95	0.37	0.90	1.00	0.17	0.38	2.65	1.13	1.09	1.43	0.04

Acceleration due to gravity (m/s²)

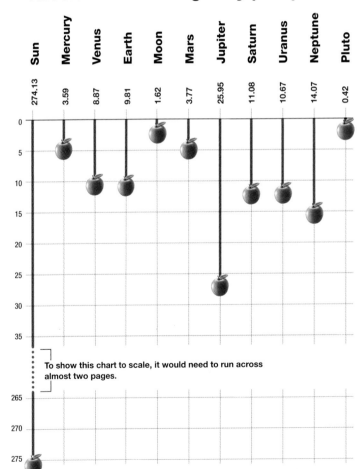

Sun	Mercury	Venus	Earth	Moon	Mars	Jupiter	Saturn	Uranus	Neptune	Pluto
274.13	3.59	8.87	9.81	1.62	3.77	25.95	11.08	10.67	14.07	0.42

To show this chart to scale, it would need to run across almost two pages.

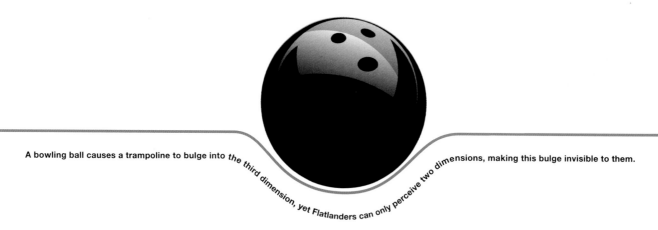

A bowling ball causes a trampoline to bulge into the third dimension, yet Flatlanders can only perceive two dimensions, making this bulge invisible to them.

Flatlanders

▶ When we try to visualize gravity as the curvature of space–time we are like Flatlanders, inhabitants of a two-dimensional world, responding to a three-dimensional mass deforming their plane.

What if gravity were not a "pull," but simply like falling down a slope in the fabric of space–time?

Flatlanders live in a world with only two dimensions, like ants marching around on a trampoline. Place a mass such as a bowling ball on their "trampoline world" and it deforms, sagging under the weight. The Flatlanders, however, with their 2-D senses, are unable to perceive this sag; they experience only its consequences. If they try to follow what they perceive as a straight-line path across the dip where the ball is sitting, they find themselves deflected in a curve, as if pulled by a strange force. In fact they are simply following a geodesic—the shortest path between two points on a curved surface—but since they are unable to perceive the curve they cannot see this. Similarly, what we perceive as gravitational force is simply acceleration due to falling down a slope—the slope in this case is in the four-dimensional fabric of space–time.

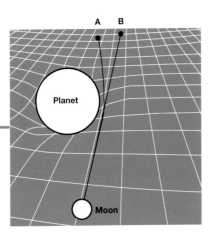

A moving object takes the shortest route; on a curved plane, such as that produced by a planet deforming space–time, this route is not a straight line (B) but a geodesic (A), a sort of curved line.

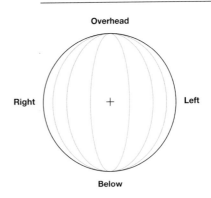

In fact there is one way to visualize the curvature of 4-D space–time from a 3-D perspective: by viewing a sphere from the inside. From within the sphere the curved longitudinal lines appear straight and parallel when looking straight ahead, even though they meet overhead.

So, for instance, Earth does not orbit the Sun because it is held by gravitational force, but because it follows the contours of space–time. In effect it is rolling round the rim of the depression in space–time made by the great mass of the Sun. The more massive an object, the more steeply it deforms space–time, and therefore the faster other objects accelerate down the slope—in other words, **making gravity stronger**.

Geodesics are familiar to us in the form of Great Circle routes. Because Earth is a sphere, the shortest distance between two points on its surface is not a straight line on a Mercator projection map (known as a Rhumb line) but a geodesic known as a Great Circle.

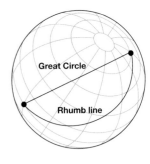

The degree by which light is bent by the Sun:

1.75 arcseconds

= 0.0004861°

Deviation of beam of light on reaching Earth =

1,269 km   789 miles

which is roughly the distance from London to Vienna. This means that starlight that would have landed on London if the Sun were not in the way actually falls on Vienna.

Rhumb line distance from New York to Hong Kong =

18,000 km   11,160 miles

Great Circle distance from New York to Hong Kong =

13,000 km   8,080 miles

5 1/2

The amount of time saved by a plane following the Great Circle route between New York and Hong Kong, as opposed to the rhumb line route: **hours**

The time-traveling twin

▶ According to special relativity, time dilation is observed by two people moving relative to one another, yet if Bill travels at 0.6 c to and from a star six light years away, when he gets back he will be four years younger than his twin Ben who stayed at home. Is this a paradox?

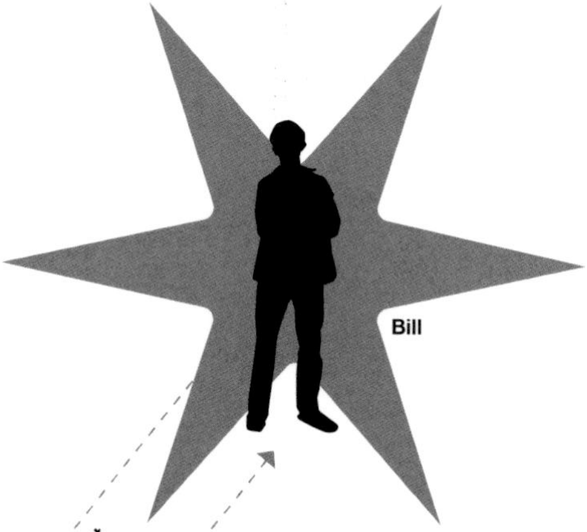

We saw from the thought experiment involving the train that observers moving relative to one another are equivalent, so that each perceives the other as in motion and therefore experiencing time dilation, yet in the so-called "twins paradox" Bill ends up younger than Ben. To see why, consider what the twins see on each other's clocks. For Ben, Bill's trip takes ten years each way, but for Bill, thanks to relativistic length contraction, the star is only 4.8 light years away and so the trip takes just eight years each way.

The twins apparently move relative to one another, so why does one of them end up younger? In fact the twins do not experience symmetrical situations; when he turns around after reaching his destination, the spacefaring twin undergoes acceleration that his brother on Earth does not.

Looking through a powerful telescope, Ben witnesses Bill's arrival at the star with the spaceship clock showing eight years, but it took six years for the light from the clock face to reach Ben's telescope, so Ben's clock now reads 16 years. Meanwhile, Bill, on arriving at the star after eight years' travel, sees the light that left Ben's clock face six years ago, showing a reading of "four years." For each twin, the other one appears to be experiencing time dilation by a factor of two.

Bill

Ben

So twenty years have passed for Ben on Earth, but only sixteen years have passed for Bill on board the fast-moving ship. Ben will have aged four more years than his brother, and the twins are no longer the same age. The two twins have not experienced symmetrical situations, so they are not equivalent. Bill has left Ben's frame of reference but then come back to it, while Ben has remained in the same frame of reference. Alternatively, we can say that Bill has undergone acceleration that Ben has not because he turned around after reaching his destination (so that relative to him the rest of the universe has undergone accelerational/gravitational time dilation—see next page). A third way of looking at it is that Bill has undergone a relativistic Doppler shift (see pp. 56–7), so that on his return journey he was effectively catching up with the light that was traveling toward Ben's telescope.

Therefore for the watching Ben, the return journey is "compressed" into a shorter time, but the same phenomenon does not happen to Bill—hence the asymmetry.

Desynchronization due to relativistic time dilation happens every time an airline pilot flies across the country while her husband stays at home, but the differences involved are infinitesimal. Space station astronauts also experience dilation relative to people on Earth.

Speed of a spaceship accelerating at

1g (9.81 m/s²)

After 1 day
3,000,000 km/h
fast enough to escape gravitational pull of Milky Way

After 1 hour
127,138 km/h
faster than Earth orbits the Sun

After 353 days
1,079,252,848.8 km/h
the speed of light

The astronauts who made up the crew of the Soviet space mission *Mir EO–3* spent almost exactly a year orbiting the Earth on board the *Mir* space station, performing over 2,000 experiments, including astronomical observations and physiology research.

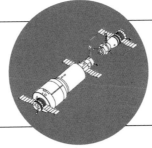

Time in orbit for: **1 year**

Speed of orbit: **8 km/s**

Time difference on touchdown: **0.01 seconds**

The old woman on top of a stepladder

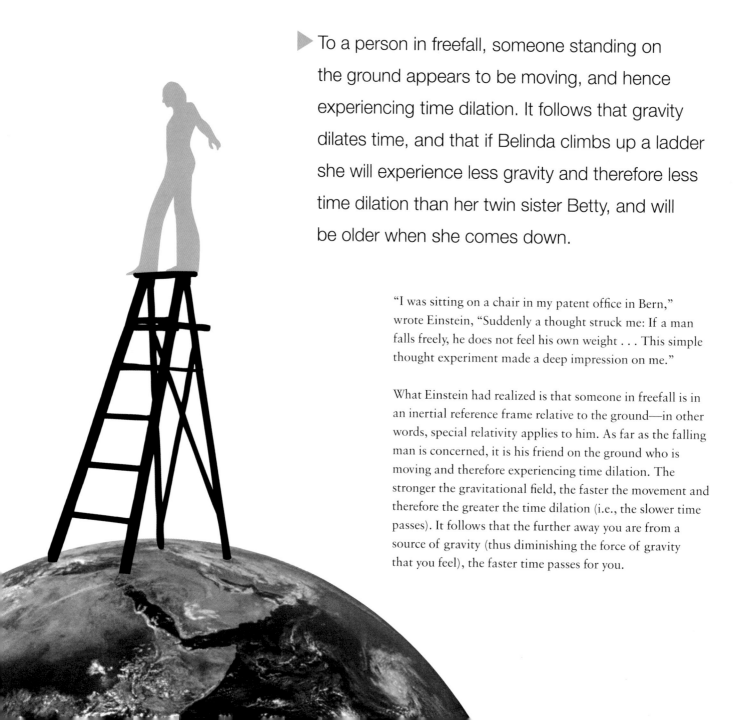

▶ To a person in freefall, someone standing on the ground appears to be moving, and hence experiencing time dilation. It follows that gravity dilates time, and that if Belinda climbs up a ladder she will experience less gravity and therefore less time dilation than her twin sister Betty, and will be older when she comes down.

"I was sitting on a chair in my patent office in Bern," wrote Einstein, "Suddenly a thought struck me: If a man falls freely, he does not feel his own weight . . . This simple thought experiment made a deep impression on me."

What Einstein had realized is that someone in freefall is in an inertial reference frame relative to the ground—in other words, special relativity applies to him. As far as the falling man is concerned, it is his friend on the ground who is moving and therefore experiencing time dilation. The stronger the gravitational field, the faster the movement and therefore the greater the time dilation (i.e., the slower time passes). It follows that the further away you are from a source of gravity (thus diminishing the force of gravity that you feel), the faster time passes for you.

Climbing a ladder moves you further away from Earth, and therefore diminishes the force of gravity acting on you, so that you will age infinitesimally faster. Following this logic, one method suggested for traveling into the future would be to fly to the surface of a neutron star, or very close to a black hole, wait there a while, and then return to Earth.

9.2

Time dilation factor at surface of neutron star. A neutron star packs into a tiny space all the mass of a much larger star, and all the gravitational attraction of a much larger star too. High gravity equals big time dilation.

The time in **minutes** experienced by an observer in space for every hour experienced by someone on a neutron star:

67

79
years

GPS satellite
Orbital height is: **20,000km**

Amount by which GPS clocks run faster due to gravitational time dilation:

45microseconds/day

Time for GPS reading to become inaccurate if not corrected for relativity:

2minutes

GPS daily error due to relativity: **10km**

Time passes more quickly at the top of the Empire State Building than at the bottom. If someone lived for 79 years at a height of 380 m (1,250 ft) they would lose 0.000104 seconds.

Mars is smaller and lighter than Earth and its gravity is about two-fifths of ours, meaning that things on Mars age faster than on Earth.

The surface of Mars is **three years** older than Earth's due to gravitational time dilation.

−0.000104seconds lost over the course of 79 years.

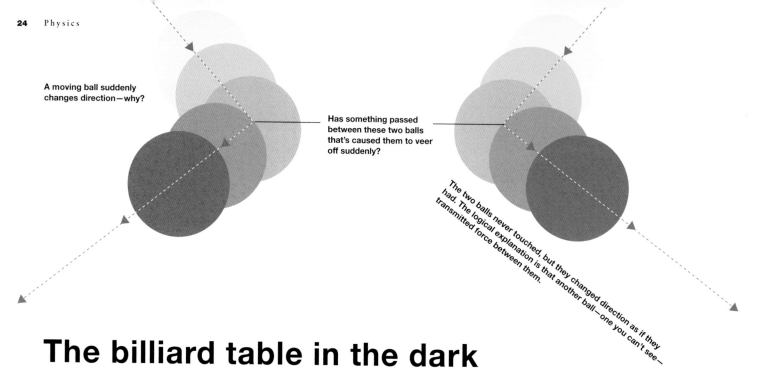

A moving ball suddenly changes direction—why?

Has something passed between these two balls that's caused them to veer off suddenly?

The two balls never touched, but they changed direction as if they had. The logical explanation is that another ball—one you can't see—transmitted force between them.

The billiard table in the dark

▶ You are looking at a billiard table in the dark. Two of the balls are luminous so you can see them approaching each other, but then they veer off without making contact—you might assume that another, non-luminous ball had passed between them, transferring force. Similarly, particle physicists deduce from the paths of particles they can see that force involves the exchange of invisible particles.

Conservation of momentum is a familiar concept often illustrated with billiard balls colliding: the sum of the forces and angles (angular momentum) of the two balls is the same before and after collision—in other words, it is conserved. Concepts of conservation of momentum underlie the analogies particle physicists use to describe how the fundamental forces of the universe—electromagnetism, strong and weak nuclear forces, and gravity—are mediated by the exchange of particles. For instance, if you saw two astronauts floating toward each other, move as if throwing and catching something, and then deviate from their previous courses, you would assume that something had passed between them, even if you could not see what it was. Equally, if you were watching two ice dancers in the dark, wearing lights on their heads, and you saw the lights approach and then revolve around one another, you would assume they had linked hands and were holding onto one another. In both cases, something physical is involved in causing the two visible particles (the astronauts or ice dancers) to either repel or attract one another; in other words to interact.

The fundamental forces compared

	Force	Strength	Range	Range equivalent
	Strong nuclear	1	0.000000000001 mm	Diameter of nucleus of a zinc atom
	Electromagnetic	0.0073	Infinite	Diameter of universe
	Weak nuclear	0.000001	0.000000000000001 mm	0.1% diameter of a proton
	Gravity	6×10^{-39} (six-thousand-trillion-trillion-trillionths as strong as the strong nuclear force)	Infinite	Diameter of universe

Similarly, force interactions between elementary particles make sense if they are understood as involving an exchange of particles, known as force carrier particles or gauge bosons. These interactions are commonly depicted using Feynman diagrams, invented by the Nobel laureate physicist Richard Feynman, which show elementary particles exchanging force carrier particles, but which could equally be used to depict any of the analogies previously described.

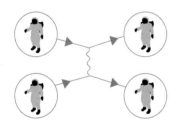

A Feynman diagram illustrating how the same principles that govern interactions between subatomic particles can apply to floating astronauts throwing wrenches to one another.

Light is made up of **photons**, which are massless. So how can they be particles? In fact Einstein's famous equation $E=mc^2$ is only a shortened version of a longer equation which says anything with momentum can be thought of as a particle.

A **human being** consists of as much energy as is found in the matter of:

30 very large H-bombs

Number of proven
elementary particles: **29**

Number of known
subatomic particles: **>200**

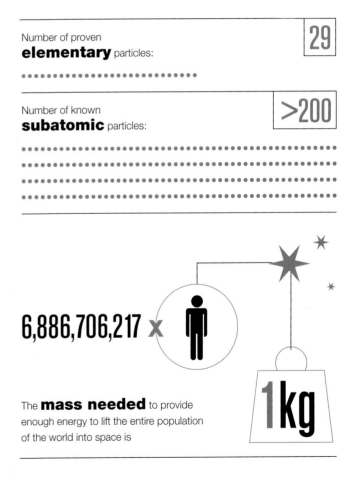

6,886,706,217 x

The **mass needed** to provide enough energy to lift the entire population of the world into space is

1kg

The string theory of everything

▶ If so-called fundamental particles are actually composed of packages of other dimensions rolled up into vibrating strings, and our universe is a membrane of four dimensions (space–time) in a megaverse of 11 dimensions, then gravity will be made of strings that span multiple extra dimensions.

The current standard model of physics, which encompasses fundamental particles and the four basic forces of the universe, has many problems. Currently, it is unable to link theories of gravity with the quantum dynamic theories that explain the other three forces. Even for these three forces, the standard model has to make all sorts of inelegant fudges, such as arbitrarily setting constants at certain levels. In their quest to find a Grand Unified Theory, or GUT, physicists have suggested that the fundamental particles and forces may not be so fundamental after all—perhaps there is

another level of structure beneath them, which could explain all their properties and bring together gravity and quantum dynamics.

One suggestion is that what we perceive as fundamental particles, such as photons and quarks, are actually the points where multidimensional strings penetrate our universe of four dimensions (aka space–time). These strings—or superstrings—vibrate, and it is these vibrations that give rise to the particles and forces we can perceive.

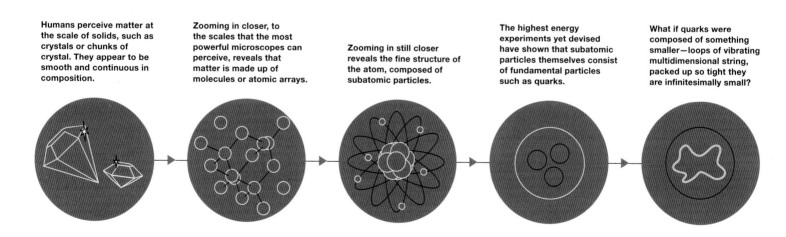

Humans perceive matter at the scale of solids, such as crystals or chunks of crystal. They appear to be smooth and continuous in composition.

Zooming in closer, to the scales that the most powerful microscopes can perceive, reveals that matter is made up of molecules or atomic arrays.

Zooming in still closer reveals the fine structure of the atom, composed of subatomic particles.

The highest energy experiments yet devised have shown that subatomic particles themselves consist of fundamental particles such as quarks.

What if quarks were composed of something smaller—loops of vibrating multidimensional string, packed up so tight they are infinitesimally small?

Big Bang timeline

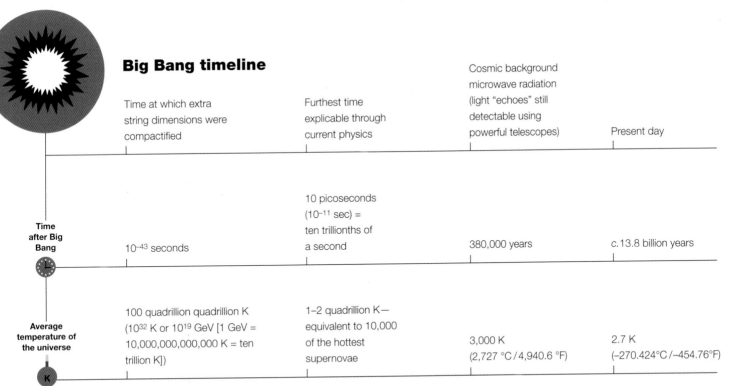

	Time at which extra string dimensions were compactified	Furthest time explicable through current physics	Cosmic background microwave radiation (light "echoes" still detectable using powerful telescopes)	Present day
Time after Big Bang	10^{-43} seconds	10 picoseconds (10^{-11} sec) = ten trillionths of a second	380,000 years	c. 13.8 billion years
Average temperature of the universe	100 quadrillion quadrillion K (10^{32} K or 10^{19} GeV [1 GeV = 10,000,000,000,000 K = ten trillion K])	1–2 quadrillion K— equivalent to 10,000 of the hottest supernovae	3,000 K (2,727 °C / 4,940.6 °F)	2.7 K (−270.424°C / −454.76°F)

In order to make the mathematics of superstring theory work, physicists have calculated that there must be ten dimensions in all, six of which are not accessible to us, possibly because they are rolled up so tightly they are imperceptible, a phenomenon known as compactification.

Alternatively, there is an 11th dimension, and our 4-D world is just one "surface" or membrane in the 11-D megaverse, just as the Flatlander's flat world or the ant's trampoline world (see pp. 18–9) is a 2-D world in a 3-D space.

Other commonly used analogies are the screen on a computer, which is a 2-D surface that is part of a 3-D whole. In this Membrane- or M-Theory, some of the superstrings are compactified and confined to our membrane (referred to as a "brane"), but some, particularly gravity strings, stretch across multiple branes. This accounts for the weakness of gravity compared to the other forces—whereas they are focused exclusively in our brane, gravity spreads its force between many branes, so that we feel only an echo of it.

If the first millisecond of the universe were slowed down to encompass the age of planet Earth, currently explicable physics would have kicked in after **50 years**, while the extra string dimensions would have been compactified just **0.15 seconds** after the formation of Earth.

100 quadrillion quadrillion K.

The almost inconceivable temperature of the universe a fragment of a second after the Big Bang. It is equivalent to a billion trillion supernovae.

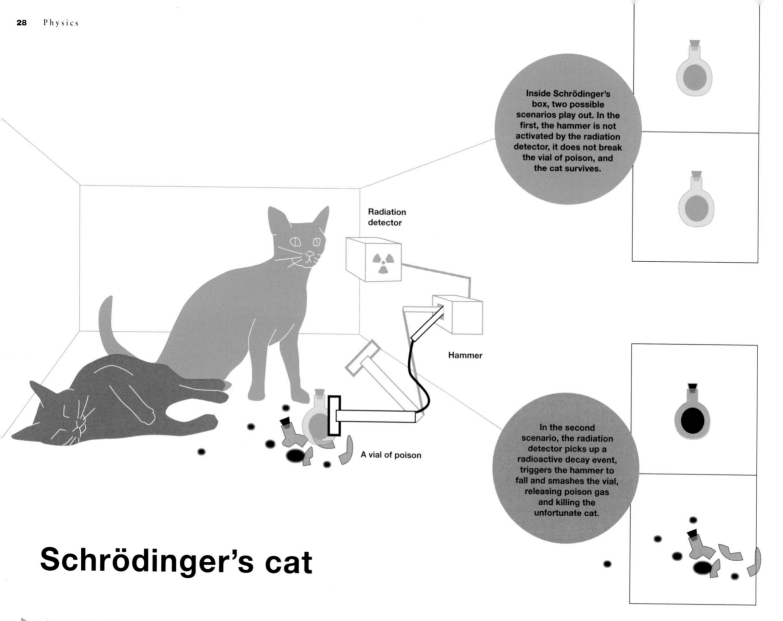

Radiation detector

Hammer

A vial of poison

Inside Schrödinger's box, two possible scenarios play out. In the first, the hammer is not activated by the radiation detector, it does not break the vial of poison, and the cat survives.

In the second scenario, the radiation detector picks up a radioactive decay event, triggers the hammer to fall and smashes the vial, releasing poison gas and killing the unfortunate cat.

Schrödinger's cat

▶ A sealed box contains a live cat and a poison-gas device triggered by a quantum event, which remains indeterminate until observed; thus according to one interpretation of quantum mechanics, until the box is opened, the cat is literally both alive and dead.

Quantum mechanics describes how particles behave like waves, which raises many paradoxes. One of these is that waves can exist in two forms at once, known as a superposition. If you look at the sea, for instance, you can see that big waves have small ripples on them; both exist at the same time, forming a superposition. When describing particles this is counterintuitive, such as when describing the radioactive decay of a particle. The particle can either decay or not decay, and normally we would say it must do one or the other. However, the mathematical principle governing this process (known as the "wave function") contains a superposition of both results, so that all that can be determined is the probability of one outcome versus the other. The only way to know for sure, according to one interpretation of quantum mechanics, is to look. This is known as the Copenhagen Interpretation.

Austrian physicist Erwin Schrödinger (1887–1961) devised a famous thought experiment to show how the Copenhagen Interpretation posed a serious challenge to traditional concepts of reality. He imagined a cat sealed into a box along with a "diabolical device": a vial of cyanide and a hammer connected to a Geiger counter. The counter was next to a radioactive particle with a 50% chance of decaying; if the particle decayed the counter would fire, triggering the hammer to fall, smashing the vial and killing the cat. According to the Copenhagen Interpretation, however, until the outcome of the process was observed, the only valid description was provided by the wave function, in which both outcomes existed at once. Until the box was opened, the cat was literally both alive and dead. Obviously, this does not match up with normal conceptions of reality, in which the hapless cat must be one or the other.

An extension of the Schrödinger's cat paradox proposed by the physicist Eugene Wigner (1902–1995)—known as **"Wigner's friend"**—pointed out that although Wigner might open the box and collapse the wave function, if he were in a sealed room then as far as his friend outside the room is concerned the superposition still exists. This "quantum indeterminacy" means that the observer's paradox could continue *ad infinitum*.

In quantum mechanics, the smallest possible distance is known as the **Planck length** (1.6 x 10^{-35} metres), and the time it would take a photon to cross that distance is known as Planck time (5.4 x 10^{-44} seconds).

X14,350,000   If you counted off one Planck length per second, it would take 14,350,000 times the current age of the universe to measure the diameter of an atom in Planck lengths.

A particle accelerator powerful enough to investigate the Planck scale would weigh as much as the Moon and have a circumference equal to the orbit of Mars.

Planck time:

0.00000000000000000000000000000000000000000054 **seconds**

The Goldilocks universe

▶ Like Goldilocks, we need a universe that is "just right" for us; if a set of fundamental constants, which govern the properties of matter and energy, were even minutely different, we (intelligent life) could never have arisen.

It is self-evident that our universe is set up to allow the existence of intelligent life, or we would not be here to make this observation, but physicists and cosmologists struggle to explain some aspects of this "setup." For instance, one factor that allows us to exist was the transformation of the hydrogen that was present in the young universe into heavier elements, such as carbon, which in turn depended on the existence of just the right ratio of neutrons to protons. This was only possible thanks to the precise ratio of the weak nuclear force to gravity.

If the weak nuclear force were slightly stronger, all of the neutrons in the young universe would have decayed and the universe would be all hydrogen. If the weak nuclear force were slightly weaker, the whole universe would be helium. There are at least five other "anthropic coincidences" ("anthropic" meaning of or relating to humans): fundamental constants that appear to have been "set" at just the right level, a phenomenon known as the "fine-tuned universe." To explain these coincidences, physicist Brandon Carter suggested there was an Anthropic Principle governing the universe: "The universe must have those properties which allow life to develop within it at some stage in its history." (Carter later regretted using the word "Anthropic," because in fact the existence of any form of intelligent observer, not just humans, validates the principle.)

The Anthropic Principle has been seized upon as an argument for the existence of God; if the universe appears to have been designed with the precise parameters for intelligent life, there must be a designer (this is a version of the teleological argument—see pp. 126–7). Other commentators dismiss the principle as a form of tautology or fallacy, essentially a version of the meaningless statement, "the universe must be set up to

"too much hydrogen"

If the weak nuclear force were stronger than it is, the universe would be a big soup of pure hydrogen.

"mmm just right"

Because the weak nuclear force is exactly as strong as it is, the universe contains a whole range of elements.

"too much helium"

If the weak nuclear force were any stronger, all non-dark matter in the universe would be in the form of helium.

The amount of time after the Big Bang when helium formation began:

180 seconds

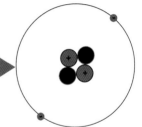

The makeup of our universe **15 minutes** after the Big Bang: **10%** helium, **90%** hydrogen

allow our existence because we exist." The principle has even been extended to a form known as the Final Anthropic Principle, which states that intelligence must arise, and once it does so can never become extinct.

Finding a perfect fit

An alternative interpretation of the Anthropic Principle is that it is evidence for a "multiple worlds" or multiverse interpretation of reality. Astronomer Martin Rees points out that you would not be surprised to find a suit that fit you exactly if you went into a department store stuffed full of suits in different sizes. Similarly, if there are many—possibly even an infinite number of—different universes, it is to be expected that at least one of them would have the parameters involved in the anthropic coincidences.

One of the fundamental constants that seems to have been set to a level favoring the existence of intelligent life is the **cosmological constant**, which explains the acceleration of the expansion of the universe. The measured size of this constant is very different from the size predicted by quantum theory.

The **discrepancy** between the measured and predicted size of cosmological constant:

10^{120}

Most of the fundamental constants have been measured with an accuracy of a few parts in a million, which corresponds to determining the length of a football field to within the thickness of two pieces of paper.

Some quantities have been measured with accuracies approaching one part in a trillion, which corresponds to determining the distance from New York to San Francisco to within one-tenth the thickness of a piece of paper.

There are an average of six hydrogen atoms in the universe per cubic meter.

3–10%

The proportion of hydrogen in a typical galaxy **today**

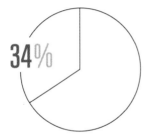

34%

The proportion of hydrogen in our galaxy **8 billion** years ago

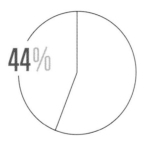

44%

The proportion of hydrogen in our galaxy **10 billion** years ago

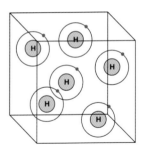

Angry bees in shrinking boxes

▶ Like a bee in a shrinking box getting angrier and expending more and more energy in an increasingly confined space, we know, thanks to the Heisenberg uncertainty principle, that an electron's velocity increases as its location is pinned down.

The Heisenberg uncertainty principle says that it is impossible to know both the location and velocity of a particle with complete certainty, and by extension that the more precisely you pin one down, the more uncertain the other becomes. In terms of velocity, increasing uncertainty means an increasingly greater range of possible velocities, including increasingly high ones. So if the location of an electron were confined to a small space, its possible velocity could become astronomical—like an angry bee in a shrinking box, the more tightly it is confined the faster it buzzes about.

$$\Delta x \, \Delta p \geq \frac{\hbar}{2}$$

A mathematical expression of the uncertainty principle, showing how the probability of a particle being at a specific location (x) and having a specific momentum (p) can never be less than a fraction of the Planck constant (h—see p. 29).

This phenomenon underlies other phenomena such as quantum tunneling, where subatomic particles can "tunnel" across barriers they should be unable to surmount. The uncertainty principle also explains why, in the atom, negatively charged electrons orbiting the positively charged nucleus are not pulled in by electrostatic attraction. If the electron were to collide with the nucleus this would constrain its location and accordingly its velocity would become too great to remain confined there. Instead, electrons orbit outside the nucleus, providing the electrostatic charge that makes the atom solid. The different orbitals or shells that the electrons occupy are constrained to discrete intervals because the electrons are waves as well as particles: they can only resonate at certain pitches and harmonics of those pitches; these correspond to the different electron shells.

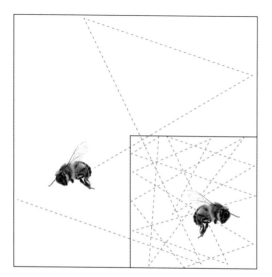

The smaller the space in which the bee is confined, the angrier it gets and the more energetically it buzzes about. An electron behaves in a similar fashion.

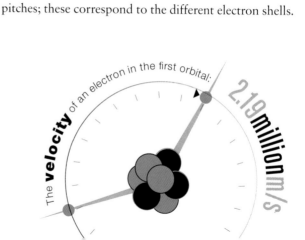

The **velocity** of an electron in the first orbital: 2.19 million m/s

The wavelike nature of a particle at the level of quantum physics has a number of counterintuitive consequences; one of the most confounding is that it allows the particle to cross impassable barriers, in a phenomenon known as **quantum tunneling**. This results from the fact that a wave does not stop abruptly—it tails off. So although the wave describing the probability of a particle being at a certain location may peak on one side of the barrier, its "tail" may extend across the barrier, making it possible for the particle to appear on the other side (see diagram below).

Classical Physics

Quantum Physics

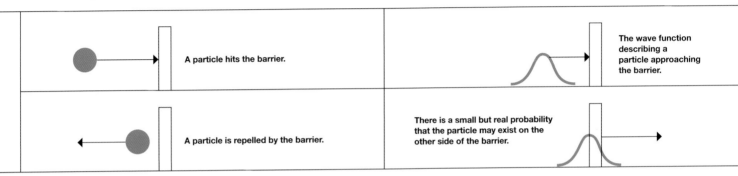

A particle hits the barrier.

The wave function describing a particle approaching the barrier.

A particle is repelled by the barrier.

There is a small but real probability that the particle may exist on the other side of the barrier.

The theoretical temperature necessary to sustain nuclear fusion in the Sun without quantum tunneling:

10 billion°C

The actual temperature of the Sun's core:

15 million°C

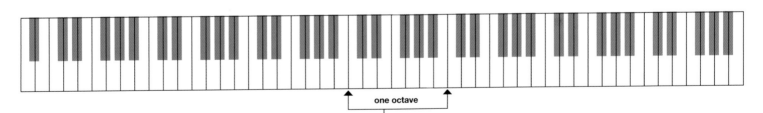

one octave

Sound waves created by a black hole in the Perseus cluster are B-flat, 57 octaves below the keys in the middle of a piano. This is a million billion times deeper than the human ear can hear. The note has been sounding for **2.5 billion years**.

57 octaves

The **biggest** pipe organ in the world is the Wanamaker Grand Court organ in Philadelphia, Pennsylvania.

It has **28,482** pipes it and weighs **287** tonnes.

Pool balls in Paris and Pittsburgh

▶ Quantum entanglement is like having two pool balls sitting side by side until a pole swings between them, setting them spinning in opposite directions, whereupon one is shipped away to Paris and the other to Pittsburgh, spinning all the while. Observing the spin of the ball in Paris instantaneously determines the spin of the one in Pittsburgh.

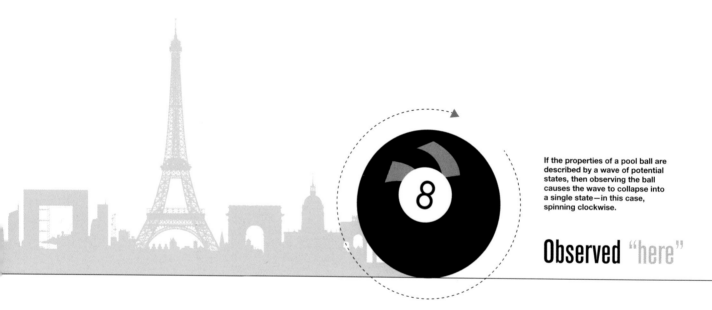

If the properties of a pool ball are described by a wave of potential states, then observing the ball causes the wave to collapse into a single state—in this case, spinning clockwise.

Observed "here"

Quantum entanglement is a counterintuitive phenomenon linked to the concept of quantum nonlocality. Fundamental particles like electrons have a property known as spin. If a pair of electrons is created together, their spins cancel each other out, so that if one is clockwise the other must be counterclockwise. But because the electrons can be described as waves, it is possible for them to have a superposition (see pp. 28–9) of the two states in which the pair is both clockwise–counterclockwise and counterclockwise–clockwise. Only when one of the pair is observed is the spin of the other electron determined, and determined instantaneously.

If the pair of electrons is separated, then each of them, like Schrödinger's cat, exists in a dual-possibility state. With the cat, looking in the box (observation/measurement) collapsed its wave function and determined an outcome, but there was only one cat. In the case of the paired electrons, observing/measuring one of them collapses the wave function for both of them, so that if electron A is taken to Paris and then has its spin measured, electron B in Pittsburgh will instantaneously acquire the opposite spin, as if the two were connected by some sort of faster-than-light telepathy. This is known as entanglement.

It is as if you set two pool balls, one black and one purple, spinning in opposite directions with a randomized device, so that without looking you knew only that there was a 50-50 chance that the black ball was spinning clockwise and the purple one counterclockwise, and then sent one to Paris and one to Pittsburgh, each in a sealed vessel. Until you opened the box in Paris and checked in which direction the black ball was spinning, the purple ball could truly be said to be spinning in both directions; but at the moment of opening the box in Paris, the Pittsburgh ball starts spinning in the opposite direction. Entanglement was actually measured in 1982 by Alain Aspect and his team at the University of Paris, who proved that entangled photons "transmitted" information faster than the speed of light.

In Aspect's entanglement experiment, the entangled photons were separated to a distance of:

13 meters

The maximum time taken for entanglement information to transfer between them:

10 nanoseconds

Distance traveled by light in 10 nanoseconds:

3 meters

Determined "there"

Since the state of the second pool ball was dependent on the state of the first, observing the first ball, and thus determining its state, also determines the state of the second ball—in this case, spinning counterclockwise.

Entanglement can be used to teleport information across space, and this information can be used to reconstitute matter from one end of the teleporter to the other.

The furthest distance of teleportation achieved to date via quantum entanglement:

404 kilometers (253 miles)

The time taken to transmit all the information needed to teleport a human, using the fastest system today:

100 million times the age of universe.

Radioactive half-life
and a handful of coins

▶ A mass of radioactive atoms with a half-life of four minutes is like a collection of coins that get tossed every four minutes.

If all the heads are flipped at fixed intervals, the number of heads will halve at the same rate; similarly, the number of atoms of a radioactive element halve at a fixed rate—its half-life.

 heads  tails

The half-life of a radioactive substance is the length of time in which, on average, half of the atoms will undergo radioactive decay. For instance, the half-life of bismuth-212 is 60.5 minutes, which means that in a sample of bismuth-212 half the atoms will have decayed after 60.5 minutes. Half of the remaining atoms will have decayed after another 60.5 minutes, and so on. Imagine you had 100 coins, all with their "heads" facing up, and you simultaneously flipped them all every 60.5 minutes. The first time you did this about half of the coins would come up tails, which equates to undergoing radioactive decay. If you then discarded the tails coins and, after another 60.5 minutes, flipped the remaining coins, you would end up with roughly a quarter of the original number facing up. Similarly, if you started off with a sample containing a million atoms of

Hydrogen is the element with the longest half-life:

10^{30} years

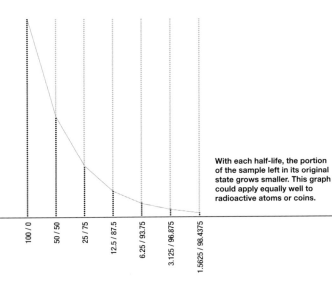

100 / 0   50 / 50   25 / 75   12.5 / 87.5   6.25 / 93.75   3.125 / 96.875   1.5625 / 98.4375

With each half-life, the portion of the sample left in its original state grows smaller. This graph could apply equally well to radioactive atoms or coins.

Tellurium-128 is the radioactive element with the longest half-life:

2.2
2.2×10^{24} years
trillion trillion years

Hydrogen-7 is the radioactive element with the shortest half-life:

2.3×10^{-23} seconds

Neutrinos are formed during beta radiation decay.

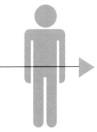

875 trillion

The number of neutrinos that pass through your body at any given moment. Neutrinos are particles formed during beta radiation decay. Because they have no charge or mass, and are only affected by the weak nuclear force, neutrinos have an astronomically small chance of interacting with matter, so they usually pass straight through.

bismuth-212, after 121 minutes (two half-lives, or 2 x 60.5) you would have roughly 250,000 bismuth atoms left. The other 75% would have decayed. How long would it take for you to end up with around 15,625 atoms of bismuth-212? Note that the number of atoms you start with is irrelevant; the half-life remains the same. If you had only one bismuth-212 atom—or only one coin—then there would be a 50% chance of it decaying after 60.5 minutes. Where the analogy breaks down is that coin-tossing only appears random: if you had all the requisite data and a super-computer to crunch it, you could probably predict whether a given coin would fall heads or tails. Radioactive decay, on the other hand, is a truly random process. All you can do is calculate a probability.

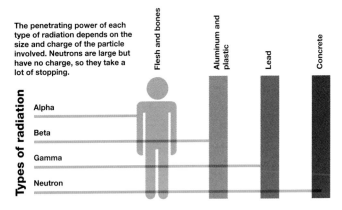

The penetrating power of each type of radiation depends on the size and charge of the particle involved. Neutrons are large but have no charge, so they take a lot of stopping.

Flesh and bones   Aluminum and plastic   Lead   Concrete

Types of radiation

Alpha

Beta

Gamma

Neutron

Heat, pressure, and pool balls

▶ A container of gas is like a box full of incredibly energetic, bouncing pool balls, each covered in little springs. Squeezing the box makes them bounce faster, just as pressurizing a gas raises its temperature.

Back in the 17th century, pioneering chemists like Robert Boyle (1627–1691) realized that they could describe the behavior of a gas if they made a number of assumptions about it, assumptions that approximated to reality but that described an "ideal" gas. An ideal gas is described as being rather like a collection of minuscule pool balls, each covered with tiny springs. Like pool balls, the particles of a gas zoom along in a straight line until they collide with one another or with the walls of their container. Thanks to their tiny springs, however, these collisions are perfectly elastic; in other words, the balls bounce off one another without losing any energy, zooming off in another direction until the next collision. Where the analogy breaks down is that while pool balls covered in springs have definite diameter and volume, the particles in an ideal gas are so tiny compared to the volume of space they occupy that they are considered point particles, essentially without dimensions.

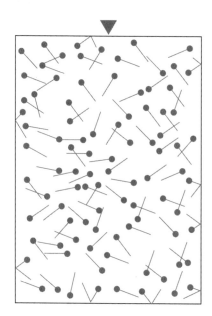

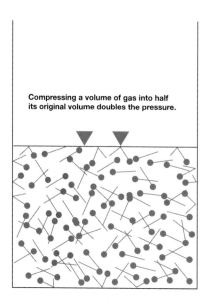

Compressing a volume of gas into half its original volume doubles the pressure.

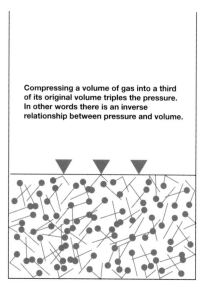

Compressing a volume of gas into a third of its original volume triples the pressure. In other words there is an inverse relationship between pressure and volume.

The average speed of molecules in the atmosphere:

~450 m/s

The average length of time between collisions:

2.36 x 10^{-7} seconds

Collision frequency:

4.23 x 10^6 per second

Radon is one of the heaviest gases:

radon	**steam**	**hydrogen**
per mass of 22.4 liters weighs **222 g**	per mass of 22.4 liters weighs **18 g**	per mass of 22.4 liters weighs **2 g**

This set of assumptions, known as kinetic theory, explains how a gas exerts pressure on its container—through the continuous battering of the walls by the billions of tiny particles—and allowed the development of simple mathematical formulas describing the relationships between temperature, pressure, and volume.

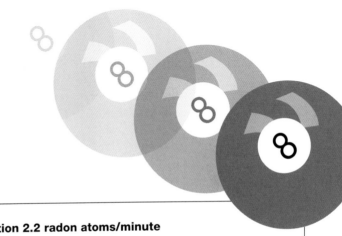

Radon, a household hazard   1 picoCurie = formation 2.2 radon atoms/minute

The average indoor level of radon gas:

1.3 picoCuries per liter

average outdoor level of radon gas:

<0.4 picoCuries per liter

1.15 million

is the number of radon atoms accumulated in a typical home in **one year**.

Mass of air needed to be exchanged to disperse this to safe levels:

5 million liters

exposure level considered to be a health hazard:

3-4 picoCuries per liter

Deaths due to lung cancer as a result of exposure to radon gas in the U.S.:

~20,000 deaths per year

Dominoes fall faster than books

▶ Metals are better heat conductors than non-metals because, just as a row of dominoes falls faster than a row of books, so too do the free electrons in metal transmit heat energy faster than the interatomic/molecular bonds.

Conduction of heat can partly be explained through simple kinetic theory, which we met in the discussion of ideal gases (see the previous entry). If you imagine a solid (a collection of particles joined together by chemical bonds) as a collection of balls joined together into a lattice by springs, conduction of heat is easy to visualize. Heat energy applied to a particle is transformed into kinetic energy, making the particle vibrate. If one of the balls in the ball-and-spring model vibrates, it will transmit its vibrations through the springs to the other balls; in the same manner, heat is conducted through transmission of kinetic energy from one vibrating particle to another.

But what makes metals more conductive than wood or other organic substances? The answer is free electrons. Free electrons are able to move from one particle to the next relatively easily. Substances such as wood do not have free electrons. They can only conduct heat through vibration of the whole atom—which, though tiny, is still much bigger than an electron alone. To see why this makes such a difference to conductivity, conduct the following experiment. Stand a row of 20 books next to a row of 20 dominoes. Now simultaneously tip over the first domino and the first book; because the dominoes are smaller they will fall faster. Similarly, in a metal the small electrons are able to move about much faster, which means they are much more efficient at transmitting energy.

Conduction in non-metals

Passing from one particle to the next, heat spreads incrementally through the solid.

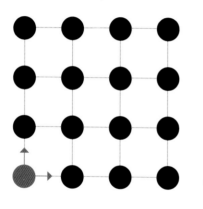

Heat, in the form of kinetic energy, is conducted along interatomic, or intermolecular bonds.

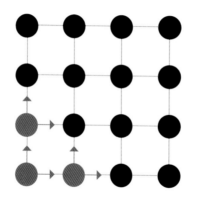

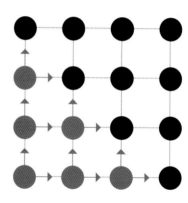

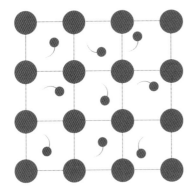

 They can conduct energy around the solid much faster than interparticle bonds alone.

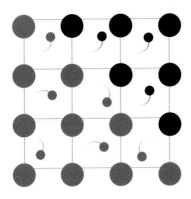

The free electrons in a metallic solid are available as energy conductors, in addition to the interparticle bonds.

Conduction in metals

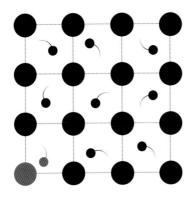

Thermal conductivity in watts per meter-Kelvin
for selected materials (at room temperature except where otherwise indicated)

Material	W/mK	Material	W/mK
Air at sea level	0.025	Neoprene	0.15–0.45
Air at 10,000 m	0.020	Nickel	90.7
Aluminum	237	Paper	0.04–0.09
Asbestos	0.05–0.15	Particle board	0.15
Brick	0.18	Plaster	0.15–0.27
Carbon (diamond)	895	Platinum	71.6
Carbon (graphite)	1950	Plywood	0.11
Carpet	0.03–0.08	Polyester	0.05
Concrete	0.05–1.50	Polystyrene foam	0.03–0.05
Copper	401	Polyurethane foam	0.02–0.03
Cotton	0.04	Sand	0.27
Feathers	0.034	Silica aerogel	0.026
Felt	0.06	Silver	429
Fiberglass	0.035	Snow (< 0°C)	0.16
Freon 12 (liquid)	0.0743	Steel (plain, 0°C)	45–65
Freon 12 (vapor)	0.00958	Steel (stainless, 0°C)	14
Glass	1.1–1.2	Straw	0.05
Gold	317	Teflon	0.25
Granite	2.2	Titanium	21.9
Helium gas	0.152	Tungsten	174
Helium I (< 4.2 Kelvins)	0.0307	Vacuum	0
Helium II (< 2.2 Kelvins)	~100,000	Water (ice, 0°C)	2.2
Iron	80.2	Water, (liquid, 0°C)	0.561
Lead	35.3	Water, (liquid, 100°C)	0.679
Limestone	1	Water, (vapor, 0°C)	0.016
Marble	1.75	Water, (vapor, 100°C)	0.025
Mercury	8.34	Wood	0.09–0.14
Mica	0.26	Wool	0.03–0.04
Mylar	~0.0001	Zinc	116

There's no "I" in team

▶ An electric circuit or a magnet is like a soccer team; although there's no contact between players, the actions of any one affects everyone on the field.

Gravity, magnetism, and electricity act at a distance; they are forces that spread through space to give fields. Since fields cannot be seen, and transfer of force via a field does not involve physical contact between particles (at least, not on scales bigger than the quantum scale—see pp. 34–5), this mechanism of action is hard to visualize. When a system such as an electrical circuit creates an electrical field, any one element of the circuit affects the whole field and is affected in turn by the whole field, but this is often counterintuitive. You might expect a battery, for instance, to supply more current to the bulb that is nearest to it than to the third one in a series; in fact, all three bulbs draw the same current.

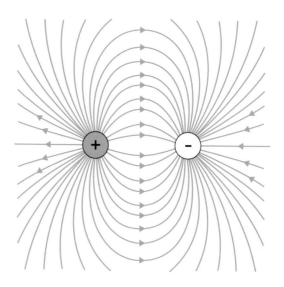

Lines of force show how the magnetic field spreads through space between the positive and negative poles of a magnet.

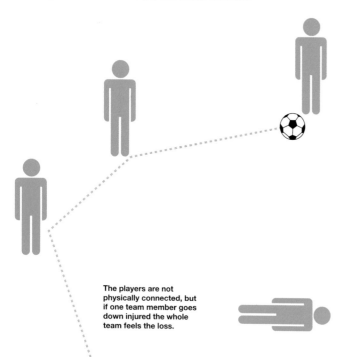

The players are not physically connected, but if one team member goes down injured the whole team feels the loss.

Similarly, a soccer team out on the playing field is made up of players who are not physically connected, but influence one another nonetheless. All are bound by the same rules of play (just as elements in a circuit obey the laws determining electric interactions) and all of them know what is going on across the whole field, even though only one of them may have the ball. More subtle influences can be at work. For instance, if the star player is injured the morale of the whole team may drop, even though they are perfectly fit. A parallel might be increasing the resistance of one stretch of wire in the circuit, which will affect the electric field intensity of the whole circuit.

The UK **national grid** has
7,200 kilometers (4,500 miles)
of overhead lines

1 kilowattperhour

One kilowatt-hour of electricity is enough to run a number of typical domestic tasks:

| 21.6 | **World** electricity production (2016):
trillion kWh |

It could light a 40-watt light bulb for 24 hours ...

Or a personal computer for two and a half hours ...

Or a 19-inch color television for four hours ...

Or a clothes dryer for a quarter of an hour ...

In human terms, one kilowatt-hour =
2,600,000 foot-pounds
(3.5 million joules or 840 kcal)

This pie chart shows that residential and commercial users are the main consumers of electricity in a developed country such as the USA.

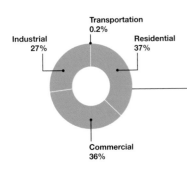

Transportation
0.2%

Industrial
27%

Residential
37%

Commercial
36%

This is enough energy to lift
907 kilograms (2,000 lbs)

a distance of
396 meters (1,300 ft)

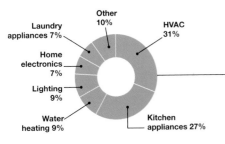

Other
10%

HVAC
31%

Laundry
appliances 7%

Home
electronics
7%

Lighting
9%

Water
heating 9%

Kitchen
appliances 27%

According to U.S. government figures for 2001, the main use of electricity in the home is for heating, ventilation, and air conditioning (HVAC), while kitchen appliances such as fridges and dishwashers use up 27% of domestic electricity.

This would be enough to carry a 40kg (90lb) backpack from sea level to the summit of
Mt Everest

An electrical circuit is like a skyscraper

▶ If an electrical circuit were a skyscraper, an elevator running exclusively to the top floor would be the battery, the height of the building would be the voltage, and the people would be charge carriers.

Electricity is a slippery thing to grasp; it cannot be seen or held, and it does not always behave in the way we intuitively expect it to. Accordingly, a host of different analogies are used to explain aspects of electricity. One of the crucial concepts is that of voltage, which is a measure of potential energy. In the skyscraper analogy, voltage is equivalent to the height of the building—a penny dropped from the top of the building would build up more speed and hit the ground harder than a penny dropped from the second floor, because it had more potential energy to start with. If the people in the building are equivalent to charge-carrying particles, then the nearest thing to a battery would be a one-way elevator that goes directly to the top floor, lifting people and thus raising their potential energy to the maximum possible in this skyscraper.

The flow of current is like the flow of people making their way through the corridors, down the stairwells, and out into the lobby, where the elevator whisks them back to the top.

Raising someone to the top of a skyscraper gives them a lot of potential energy; this is converted into other forms of energy as they make their way back down to the first floor. Similarly, in an electrical circuit, electrons are raised in potential energy, and convert that energy into other forms as they make their way back down the voltage gradient as they move around the circuit.

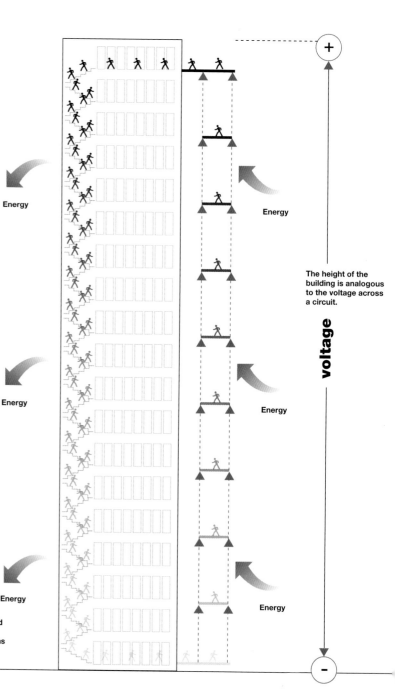

Energy

Energy

Energy

Energy

Energy

Energy

The height of the building is analogous to the voltage across a circuit.

voltage

+

−

A measure of the current could be obtained by counting the number of people passing a specified point every minute (or any other unit of time). The corridors on each floor are like wires in the circuit, and traveling along them effectively requires no energy. But when the people are walking down stairs, they convert their potential energy into heat energy, so a stairwell is equivalent to a resistor of some sort. The narrower the stairwell, the higher the resistance, since it makes climbing down the stairs harder. A light bulb is a type of resistor—just as the people walking down the stairwell convert their potential energy into heat energy, so too do charge carriers moving through the filament of the bulb convert their potential energy into heat energy (and light).

The number of light bulbs in the world:
12 billion

The number of light bulbs purchased in the USA in 2009:
2.5 billion

The 4-watt Livermore Light in Firestation 6, Livermore, California, is arguably the **longest-burning** light bulb in the world. It has been in continuous use since

1901

Total daily sales of light bulbs in the USA:
5.5 million

Annual U.S. expenditure on light bulbs:
1 billiondollars

The total amount of **energy used** by the Livermore Light during its life so far:

>13.75 gigajoules

This is equivalent to running four tumble dryers continuously for **one year**.

Conveyor belts and coal piles

▶ Current in an electrical circuit is like a conveyor belt carrying fuel from a coal pile to a furnace; although energy is delivered and used up, current itself, like the conveyor belt, is conserved.

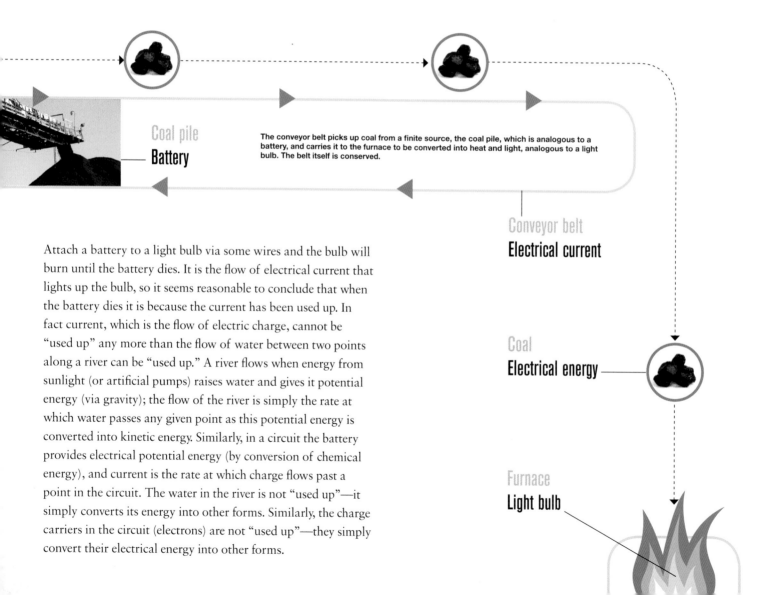

Coal pile
Battery

The conveyor belt picks up coal from a finite source, the coal pile, which is analogous to a battery, and carries it to the furnace to be converted into heat and light, analogous to a light bulb. The belt itself is conserved.

Conveyor belt
Electrical current

Coal
Electrical energy

Furnace
Light bulb

Attach a battery to a light bulb via some wires and the bulb will burn until the battery dies. It is the flow of electrical current that lights up the bulb, so it seems reasonable to conclude that when the battery dies it is because the current has been used up. In fact current, which is the flow of electric charge, cannot be "used up" any more than the flow of water between two points along a river can be "used up." A river flows when energy from sunlight (or artificial pumps) raises water and gives it potential energy (via gravity); the flow of the river is simply the rate at which water passes any given point as this potential energy is converted into kinetic energy. Similarly, in a circuit the battery provides electrical potential energy (by conversion of chemical energy), and current is the rate at which charge flows past a point in the circuit. The water in the river is not "used up"—it simply converts its energy into other forms. Similarly, the charge carriers in the circuit (electrons) are not "used up"—they simply convert their electrical energy into other forms.

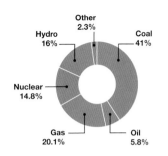

Simpler analogies indicating the conservation of current include the conveyor belt and the coal pile. The coal pile is like a battery—a store of chemical energy. The conveyor belt simply carries the fuel to the furnace (equivalent to a light bulb) where it is converted into other forms of energy; the belt itself is conserved. As long as there is coal to transfer, it keeps running. When the pile is finished it stops, but it can restart any time the coal pile is replenished. The rate at which coal is carried past a certain point is equivalent to the current in the electrical circuit. Another analogy is a bicycle chain, which transfers energy from the pedals to the wheels but is not used up.

41%
of global electricity generation is coal-powered.

At current extraction rates, the world's coal reserves will last another **110 years**

Remarkably, this is roughly **1.5 times** more than the entire current global biomass, which includes all living biological organisms anywhere in the world.

It is predicted that global **coal reserves** amount to some

1.1 trillion tonnes

7 billion tonnes

In 2009 world coal production reached nearly 7 billion tonnes—more than a tonne each for every man, woman, and child on the planet.

You could use your tonne of coal to power an old car for two and a half years, or run a tumble dryer for nine years.

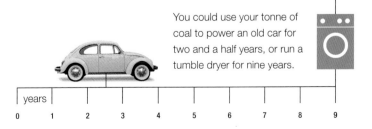

years
0 1 2 3 4 5 6 7 8 9

In a 1 mm-thick slice of typical household lighting-circuit copper wire there are as many as

3.51 x 10²⁰

3.51×10^{20}

copper atoms. To give you a sense of the magnitude of this number, it is roughly the same as the number of grains of sand on Earth.

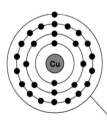

Electricity is like water

▶ An electrical circuit is like an aquarium system, with a battery for a pump, wires for pipes, and a resistor as a filter.

Because electricity flows around a circuit, water is a natural analogy to use in describing it. In a water circuit, such as that used in setting up an aquarium, the volume of water is constant, but energy is put into the system to move the water around. Similarly, in an electrical circuit the number of electrons is constant, but energy is added to the system to move them about. Just as a pump pushes the water around the aquarium circuit by building up pressure, thus giving the water potential energy, a battery builds electrical "pressure" in the form of voltage; it is this "pressure" that drives the circuit.

Water moving through pipes experiences a little resistance, but not much, so it is able to flow freely. Similarly, highly conductive copper wires offer low resistance and electrical current is able to flow freely through them. The filter in the aquarium resists the flow of water, however, just as a resistor such as a light bulb filament resists the flow of electricity. Water must do work (i.e., expend energy) to get through the filter—if you had a sensitive enough thermometer you would be able to measure a slight temperature increase as the kinetic energy of the water is converted into heat energy.

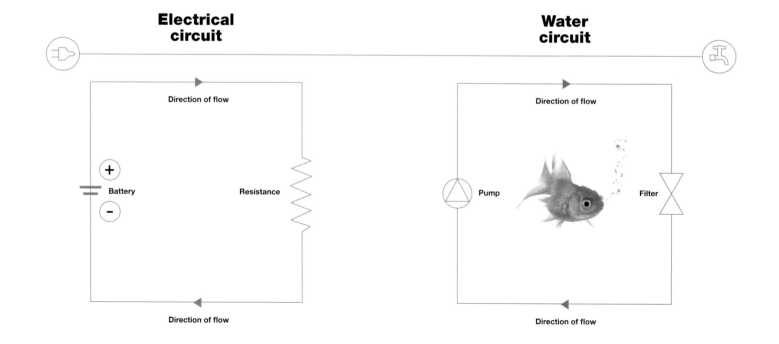

Electrical circuit

Direction of flow

Battery

Resistance

Direction of flow

Water circuit

Direction of flow

Pump

Filter

Direction of flow

Similarly, the electrical energy of the current is converted into heat and light in the bulb. The amount of water flowing through any given part of the circuit at any one time is the current, and the same applies in the electrical circuit.

The two systems are also similar in that individual water molecules and electrons might take a relatively long time to complete a circuit, but because the pipes and wires are already filled with water and electrons respectively, pushing at one end instantaneously displaces water or electrons at the other end.

1

Current is measured in amps
(charge units going past per second)

amp ~ 6milliontrillion

electrons per second

A typical car battery can supply up to 80 amps in an hour, which is

1.728×10^{24} **electrons**

or nearly 2 trillion trillion—equivalent to roughly 10,000 times the number of grains of sand on Earth.

The speed of transmission of an electrical current:

~ speedoflight

1

The speed of individual electrons in a typical circuit:

meterperhour

The percentage of energy a typical incandescent light bulb wastes as heat:

95%

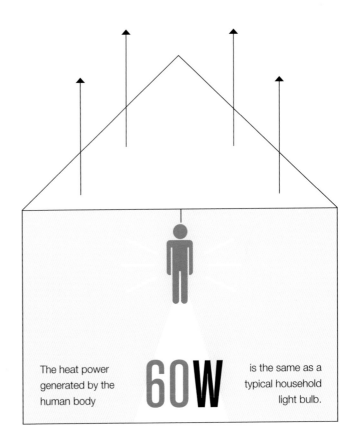

The heat power generated by the human body **60W** is the same as a typical household light bulb.

A parallel circuit is like a school hall

▶ A school hall with two exits can empty twice as fast as a hall with one exit, just as a battery connected to two light bulbs in parallel will die twice as fast as in a circuit with one bulb.

One of the most confusing and counterintuitive features of an electrical circuit is the difference between series and parallel circuits. If a circuit consists of a single wire going from a battery to a bulb, to another bulb, so that the bulbs are in a line, they are said to be in series. But if the battery is connected to one bulb with one wire and the other bulb with another wire, the bulbs are said to be in parallel. In a parallel circuit, each "loop" has the same voltage, whereas in the series circuit the same current flows through all of the bulbs so the voltage is "shared" between them.

A simple analogy for a series circuit is a garden hose fitted with multiple sprinklers. The pressure of water supplied at the start of the hose does not change, but the more sprinklers that are added, the more this pressure is shared out and the lower each jet of water becomes. Similarly, in the series circuit the light bulbs have to share the voltage, and the more bulbs that are added, the dimmer they will each glow.

The garden hose analogy breaks down for parallel circuits. In a parallel circuit, each bulb, powered by the same voltage, will glow equally brightly. A better analogy is with a school hall packed with students who want to get out. Each exit allows students to pass at a certain rate, equivalent to current flowing through a light bulb. Opening another exit will thus double the rate at which the students can leave, equivalent to adding a light bulb in parallel, which doubles the current. However, if the second doorway is farther down the corridor from the first, then the rate at which students are leaving the hall will not change—this is equivalent to a bulb added in series, which does not change the current. One obvious consequence is that, in the former scenario, the hall will empty twice as fast as in the latter; similarly, the battery in the parallel circuit will die twice as fast as the one in the series circuit.

The bulbs in the parallel circuit on the right will burn more brightly than those in the series circuit on the left, although the battery will die more quickly in the former.

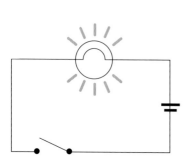

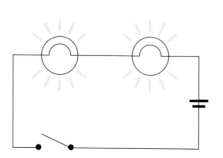

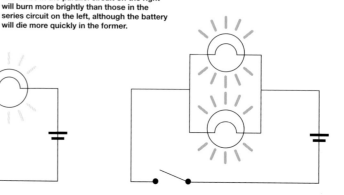

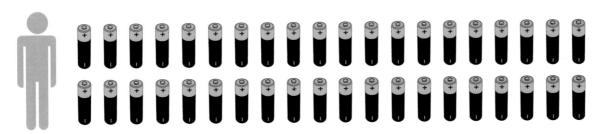

The average consumer uses **30–50** disposable batteries a year.

Worldwide, more than

15 billion batteries

are thrown away each year—enough to make a column that would stretch to the Moon and back.

In the USA alone

2.9 billion batteries

are thrown away each year.

The **world's biggest battery** is the Golden Valley Electric Association high performance nickel–cadmium storage battery in Fairbanks, Alaska.

2,000m²

Total size: 2,000m² (21,520 square feet)—bigger than a football field.

Energy cells **x13,760**

Power output:

27 MW

for up to 15 minutes.

total weight

1,300 tonnes

22 thousand tonnes

In the UK, around 600 million household batteries (equivalent in weight to 110 jumbo jets) are sent to landfills every year. The average household uses 21 batteries a year.

The world's smallest battery is a polysiloxane polymer neurological implant battery.

Dimensions: 2.9 mm in diameter and 13 mm in length (about the size of a pencil tip)—roughly ¹/₃₅ the size of a standard AA battery.

AA

x35

Matchbooks, mousetraps, and nuclear fission

▶ In a nuclear reactor or an atom bomb, a chain reaction sustains nuclear fission, in the same way that lighting one match in a matchbook, or setting off one mousetrap in a room full of traps, triggers a chain reaction.

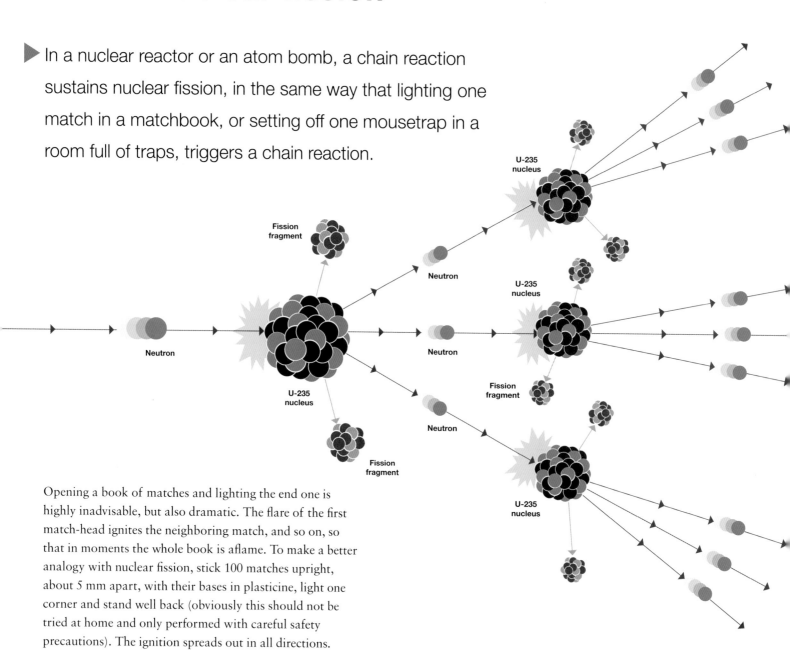

Fission fragment

U-235 nucleus

Neutron

U-235 nucleus

Neutron

U-235 nucleus

Fission fragment

Neutron

Fission fragment

U-235 nucleus

Opening a book of matches and lighting the end one is highly inadvisable, but also dramatic. The flare of the first match-head ignites the neighboring match, and so on, so that in moments the whole book is aflame. To make a better analogy with nuclear fission, stick 100 matches upright, about 5 mm apart, with their bases in plasticine, light one corner and stand well back (obviously this should not be tried at home and only performed with careful safety precautions). The ignition spreads out in all directions.

An even better model is provided by setting several dozen mousetraps and placing them side by side in an aquarium, each with three small paper balls placed on the spring-loaded trap. Toss in a ball of paper and a mousetrap will be set off, scattering its paper balls onto neighboring traps, which will be set off in turn. The mousetraps here represent atoms of uranium-235, a highly unstable radioactive isotope of uranium that can be triggered, through neutron bombardment, to decay into an atom of krypton-91, an atom of barium-142 and three high-speed neutrons (represented by the paper balls in the model). These three neutrons in turn hurtle outward; if there are other atoms of U-235 in the immediate vicinity, each of them may trigger a fresh radioactive decay event, setting off a cascade of neutron production and energetic atomic fission. In a nuclear reactor, U-238 is enriched to contain 3–4% U-235—just enough to ensure that, on average, one out of every three neutrons released collides with another atom of U-235, sustaining the fission reaction but not allowing it to run out of control. An atomic bomb uses 90% U-235, so that an explosive chain reaction is triggered.

40 million kilowatt-hours

One tonne of natural uranium produces so much energy that it is equivalent to more than

of electricity.

This is equivalent to burning

16,000 tonnes of coal

or

80,000 barrels of oil

Theoretically, a **single kilogram** of uranium-235 could produce around 80 terajoules of energy, which is equivalent to the energy that could be produced by 3,000 tonnes of coal.

In the USA, nuclear waste from reactors mostly takes the form of depleted uranium hexafluoride, known as DUF6.

The U.S. Department of Energy has accumulated **57,634** cylinders of DUF6, with a total weight greater than all eight of the U.S. Navy's Nimitz-class aircraft carriers combined, or more than 70 Ticonderoga-class cruisers. Stacked end to end, these cylinders would make a tower 219 km (136 miles) high.

U-238 decays by emitting an alpha particle

The speed of an alpha particle:

25,000 kilometers per second

100,000 X

faster than a passenger jet

Skating on rough ice

▶ Just as a skater who crosses from a patch of smooth ice onto rough ice will deviate from his course because one skate will slow down before the other, so too will a ray of light crossing from one medium to another bend or refract.

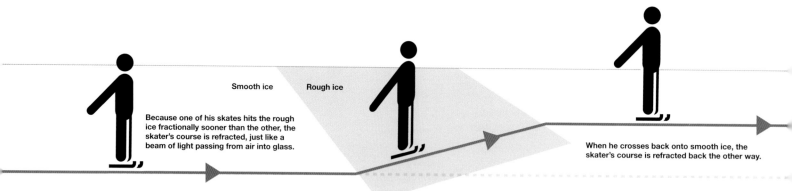

Smooth ice   Rough ice

Because one of his skates hits the rough ice fractionally sooner than the other, the skater's course is refracted, just like a beam of light passing from air into glass.

When he crosses back onto smooth ice, the skater's course is refracted back the other way.

Refraction is a phenomenon in which a ray of light bends on passing between media of different densities. When a ray of light passes from a less dense to a more dense transparent medium, it bends toward the normal (a line perpendicular to the interface between the media); when it passes from more to less dense, it bends away from the normal.

Refraction is the result of light slowing down or speeding up (the speed of light, c, is only an absolute in a vacuum); in a dense medium it slows down. The ratio between c and the speed at which light travels in a material is called the refractive index of the material. For visible light, the refractive index of glass is typically around 1.5, so that light in glass travels at $c/1.5$, which is roughly 200,000 km/s (125,000 miles/s). The refractive index of air is about 1.0003, so light travels through air at very close to c. Thus, on passing from air into a block of glass, light slows down by around a third. Why should this make it bend?

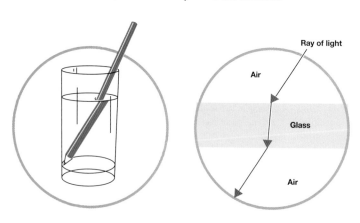

Ray of light

Air

Glass

Air

Refraction at media boundaries explains why the pencil appears distorted below the waterline—the light reflected back from the lower part of the pencil is refracted to one side by its passage through the water.

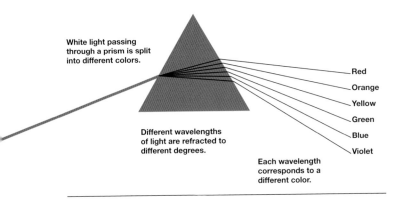

White light passing through a prism is split into different colors.

Different wavelengths of light are refracted to different degrees.

Each wavelength corresponds to a different color.

- Red
- Orange
- Yellow
- Green
- Blue
- Violet

Speed of light facts

The speed of light is affected by the medium it is passing through; the denser the medium the more it is slowed. For instance, light traveling through diamond, a very dense medium, is slowed by a factor of 2.417. The familiar term "speed of light" (c) refers exclusively to light in a vacuum. Traveling at c, light covers great distances.

Distance	Time
One foot	1.0 ns
One meter	3.3 ns
One kilometer	3.3 µs
One statute mile	5.4 µs
From geostationary orbit to Earth	119 ms
The length of Earth's equator	134 ms
From the Moon to Earth	1.3 s
From the Sun to Earth (1 AU)	8.3 min
One parsec	3.26 years
From Proxima Centauri to Earth	4.2 years
From Alpha Centauri to Earth	4.37 years
Across the Milky Way	100,000 years
From the Andromeda galaxy to Earth	2.5 million years

(★) the Canis Major Dwarf galaxy

Imagine an ice skater traveling with his legs slightly apart and his skates parallel. He crosses the dividing line between a patch of smooth ice, on which he moves faster, to rough ice, which slows him down. If he hits the dividing line straight on (perpendicular to the line, or on the normal) he will simply slow down but keep going straight on. However, if he hits it at an angle, one skate (say the left) will cross onto the rough ice a fraction of a second before the other. During this instant the left skate will be traveling slower than the right one, so the skater's right leg will travel a tiny bit further and he will rotate slightly around his center, causing him to deviate from his course—he has been refracted. Similarly, a ray of light, though very narrow, still has a degree of width, so that if the ray hits a block of glass at an angle, one edge will hit the glass first and will slow down before the other edge, causing the ray to bend.

Speed of light when traveling through a Bose-Einstein condensate of rubidium cooled to almost absolute zero

metersper second  **~60 km/h**

Shifting sirens

▶ A distant galaxy, hurtling away from our own, is like an ambulance speeding past; just as the sirens of the ambulance are Doppler-shifted to a lower pitch, the light from the galaxy is red-shifted to a longer wavelength.

The Doppler effect, named for the Austrian mathematician and physicist, Christian Doppler (1803–53), is a phenomenon caused by the relative motion of a source of waves. Sound and light are both forms of wave, so both can be Doppler-shifted, as can waves in water or any other medium. Imagine that two frogs on either side of a pond are watching a bug on the water in the center of the pond, which is twitching its legs and generating ripples (tiny waves) at a rate of four per second. If the bug is stationary, each frog sees ripples arriving at its side of the pond at a frequency of four per second.

However, if the bug starts moving toward Frog A and away from Frog B, still emitting four ripples per second, it will start to catch up the ripples ahead of it and outpace the ones behind it. The ripples in front of it will begin to "pile up," and so have smaller intervals between them; their frequency will increase and their wavelength will decrease. Frog A will see ripples arriving at a frequency higher than four per second, while Frog B will see fewer ripples per second. When an ambulance comes toward you, the sound waves emitted by the siren, like the bug's ripples, are piling up.

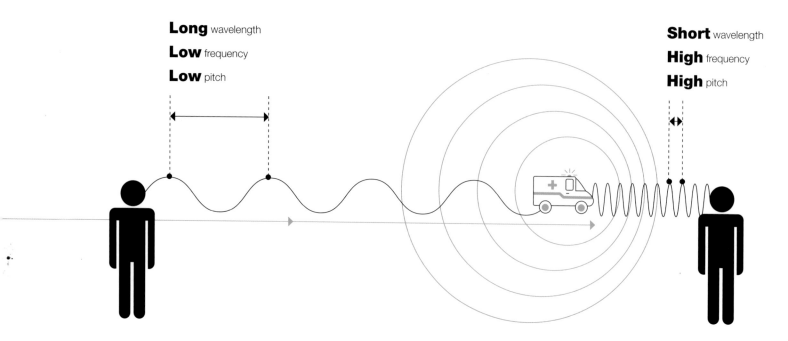

Long wavelength
Low frequency
Low pitch

Short wavelength
High frequency
High pitch

As a result, you hear sound waves of a higher frequency than the ones actually emitted by the siren. Higher-frequency sound is of a higher pitch, so you would hear the siren as a higher pitch than if the ambulance were stationary. Once the ambulance is receding from you, the sound you hear is of a lower frequency, hence a lower pitch. When a galaxy is moving away from Earth, the light it gives off is Doppler-shifted so that its frequency becomes smaller and its wavelength longer. Red light has a longer wavelength than blue light, so the light moves toward the red end of the spectrum, and is said to be red-shifted. If the galaxy is approaching Earth, the light becomes a higher frequency and shorter wavelength, and is said to be blue-shifted. Astronomers can use red and blue shifts to calculate the speed of distant objects relative to Earth. In fact, it turns out that the further away an object is, the more it is red-shifted—i.e., the faster it is moving. This is striking evidence that the whole universe is expanding, and a key plank in the Big Bang theory.

Blue-shifted   **Red-shifted**

One of the few galaxies that is moving toward us (the light of which is therefore blue-shifted) is the **Andromeda galaxy**.

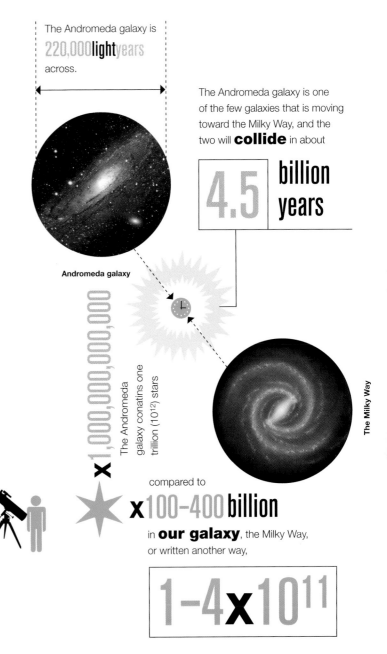

The Andromeda galaxy is **220,000 light**years across.

Andromeda galaxy

The Andromeda galaxy is one of the few galaxies that is moving toward the Milky Way, and the two will **collide** in about

4.5 billion years

The Andromeda galaxy conatins one trillion (10^{12}) stars

x1,000,000,000,000

compared to

x100–400 billion

in **our galaxy**, the Milky Way, or written another way,

1–4 x 10^{11}

The Milky Way

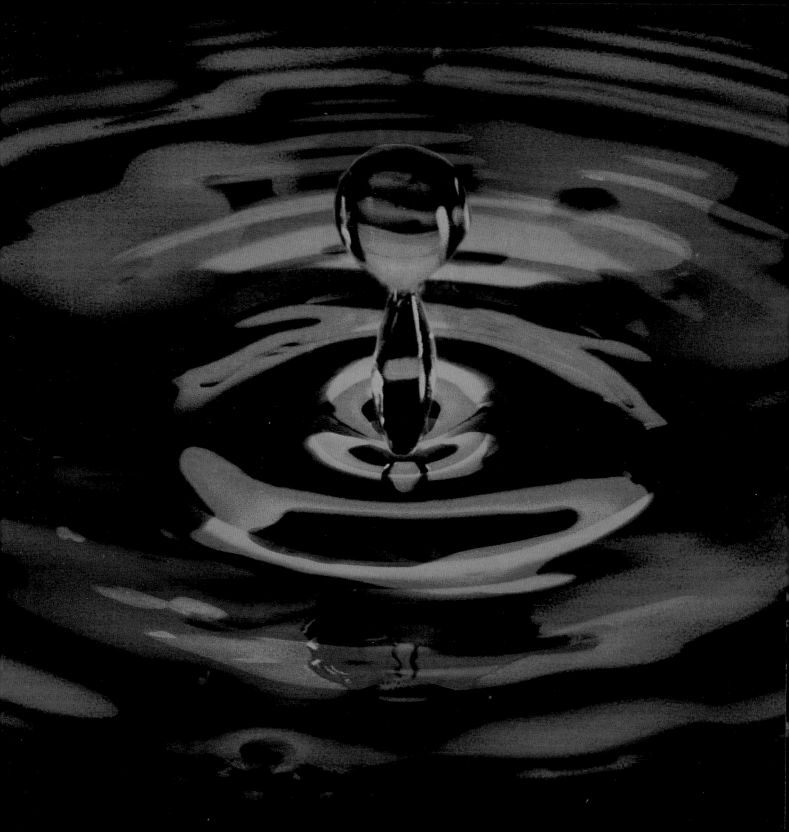

Section Two

▶ Because the essence of all chemistry takes place on a microscopic level, where atoms and molecules combine at a scale that cannot be directly observed, analogies are an invaluable tool. This section uses analogy to explain such complex subjects as atomic theory, DNA, and the behavior of powerful acids.

Chemistry

Dancing on the head of a pin

▶ Atoms are so incredibly minute that 5 trillion can fit on the head of a pin.

Medieval philosophers were lampooned for supposedly asking how many angels could dance on the tip of a needle. This exact question wasn't actually posed, but St Thomas Aquinas did debate whether angels were material beings; if they were not, presumably an infinite number could dance on the tip of a needle. Over time, this legendary debate became reworded to apply to the head of a pin, and this is now the ultimate yardstick for measuring the extremely small.

Few things are smaller than the atom. It is hard to overstate the smallness of atoms. A typical atom is 0.32 nanometers wide, which is 0.00000032 mm or 3.2×10^{-10} m. (The word "nanometer," by the way, derives from the Greek *nanos*, meaning "dwarf.") It is the electrons, which orbit the nucleus at a great distance, that give an atom its diameter.

The smallest atom is hydrogen, which has just one electron, and a diameter of 0.24 nm. Atoms do not range much in size, because although heavier elements have many more electrons, they also have many more protons in the nucleus, so the attractive force between positive and negative particles is stronger and the electrons do not venture as far. A plutonium atom weighs more than 200 times as much as a hydrogen atom, but the diameter of a plutonium atom is only about three times that of a hydrogen atom.

Hydrogen

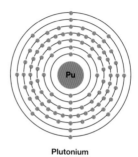

Plutonium

An atom of plutonium has 94 electrons compared to hydrogen's single electron, but because there are also 94 protons in its nucleus, the electrons are pulled in close, so that the diameter of the atom is only around three times greater.

X (5 trillion)

5,000,000,000,000

atoms on a surface of no more than a couple of millimeters across.

Sizes of Some Atoms:

Atom	Radius	
Hydrogen	0.12 NM	○
Oxygen	0.14 NM	○
Nitrogen	0.15 NM	○
Carbon	0.16 NM	○
Sulfur	0.185 NM	○
Phosphorus	0.19 NM	○

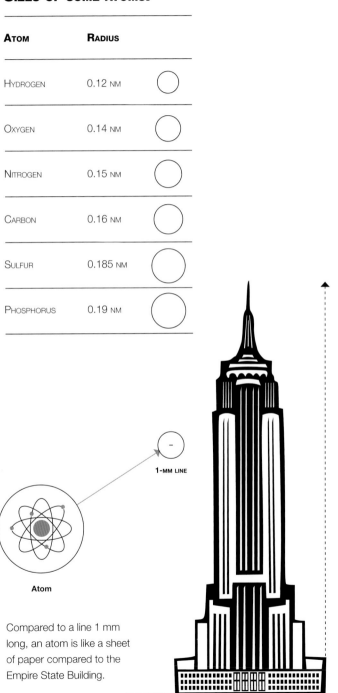

1-MM LINE

Atom

Compared to a line 1 mm long, an atom is like a sheet of paper compared to the Empire State Building.

The **atomic world** is many orders of magnitude smaller than humans are able to comprehend—even the smallest objects visible to us consist of trillions of atoms.

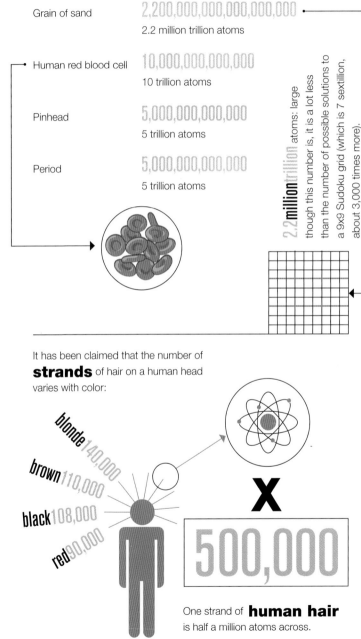

Grain of sand	2,200,000,000,000,000,000	
	2.2 million trillion atoms	
Human red blood cell	10,000,000,000,000	
	10 trillion atoms	
Pinhead	5,000,000,000,000	
	5 trillion atoms	
Period	5,000,000,000,000	
	5 trillion atoms	

2.2 million trillion atoms: large though this number is, it is a lot less than the number of possible solutions to a 9x9 Sudoku grid (which is 7 sextillion, about 3,000 times more).

It has been claimed that the number of **strands** of hair on a human head varies with color:

blonde 140,000
brown 110,000
black 108,000
red 90,000

X

500,000

One strand of **human hair** is half a million atoms across.

If an apple were as big as Earth

▶ If an apple were enlarged to the size of Earth, a hydrogen atom on the same scale would be the size of an apple.

As we saw on the preceding pages, atoms are so small it is hard to visualize them. If we cannot shrink our mind's eye down to the atomic scale, however, perhaps instead we can expand the atomic scale in our imagination. Imagine, for instance, that you took a soccer ball and blew it up to the size of Earth. An atom would be about the size of a large pea, and although you would be able to see a slight difference in sizes between the smallest atom, hydrogen, and larger ones, even an atom of plutonium would still only be about the size of a golf ball.

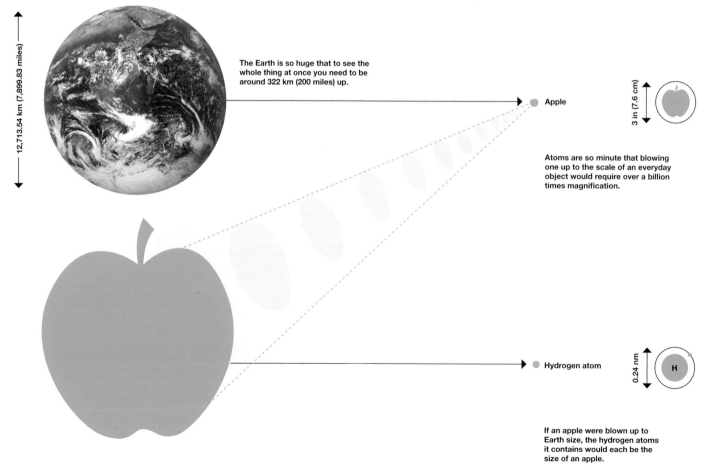

12,713.54 km (7,899.83 miles)

The Earth is so huge that to see the whole thing at once you need to be around 322 km (200 miles) up.

Apple

3 in (7.6 cm)

Atoms are so minute that blowing one up to the scale of an everyday object would require over a billion times magnification.

Hydrogen atom

0.24 nm

H

If an apple were blown up to Earth size, the hydrogen atoms it contains would each be the size of an apple.

If you wanted to play **soccer** with an atom, you would have to blow it up 1.5 billion times.

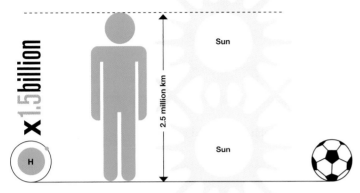

If you were enlarged to the same scale, you would be nearly 2.5 million km (1.5 million miles) high—almost the height of two Suns stacked on top of one another—and you would weigh more than the populations of China and India combined.

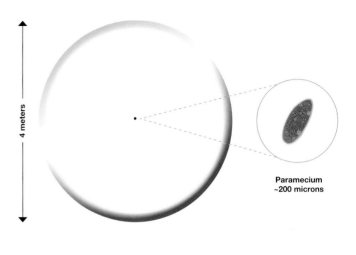

Paramecium ~200 microns

A **paramecium** is a tiny single-celled animal. If you wanted to see, swimming in a drop of water, a paramecium roughly the size of your hand, you would need to blow up a 6mm drop to be around 4 m (13 ft) across. If you wanted to see an atom with the naked eye, you would need to blow up the drop until it was 6 km (3.75 miles) wide, at which point the paramecium would be somewhere in the order of 200 m long, roughly the length of two gridiron fields.

If you inflated an atom to the size of a **basketball**, then a coin would be as big as Earth.

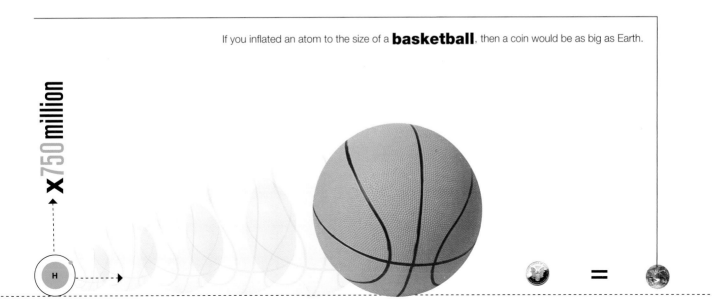

A dash of Shakespeare and a pinch of Genghis Khan

▶ Atoms are virtually indestructible, which means they hang around for ages—a billion of the atoms in your body used to be part of William Shakespeare, and a billion more belonged to Genghis Khan.

The word "atom" comes from the ancient Greek *atomos*, meaning "indivisible," "uncuttable," or "indestructible." The atom was believed to be the smallest particle of matter it was possible to get—if you kept cutting something in half, eventually you would end up with a particle that could not be cut any more, and this would be the atom.

In the 18th century chemists began to formulate the modern concept of the atom, again believing it to be indivisible, but that was not to last. In 1897 the physicist J. J. Thomson discovered the electron, a subatomic particle, and since then around 300 subatomic particles have been discovered—so many that physicists call it the "particle zoo."

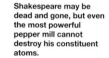

Shakespeare may be dead and gone, but even the most powerful pepper mill cannot destroy his constituent atoms.

Out of every **200 atoms** in the human body:

- 126 are hydrogen (H)
- 51 are oxygen (O)
- 19 are carbon (C)
- Three are nitrogen (N)
- The last one is split between all the other elements

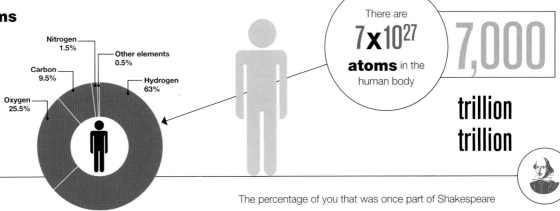

Nitrogen
1.5%

Other elements
0.5%

Carbon
9.5%

Hydrogen
63%

Oxygen
25.5%

There are
7×10^{27}
atoms in the human body

7,000
trillion
trillion

The percentage of you that was once part of Shakespeare

~ **0.00000000000000001%**

As well as discovering that the atom can be broken into smaller pieces, however, scientists have also discovered that it is incredibly difficult to do so. For most non-radioactive elements, doing any more damage than stripping off a few electrons takes enormous amounts of energy. All the elements heavier than hydrogen and helium were forged in dying stars and have probably survived passing through more than one star since then. Most of the atoms on Earth, including the ones from which you are made, have been around for billions of years, and it is statistically likely that a handful (figuratively speaking) of your component parts once made up any historical figure you care to name previous to the 20th century.

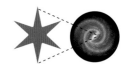

There are around **10,000** times more atoms in the human body than there are stars in the universe.

The **rarest** naturally occurring element is **astatine**; at any one time there is less than a gram of it on Earth.

Even more durable than atoms are protons, subatomic particles found in atomic nuclei. Theoretically, protons can break down into exotic particles such as positrons and neutrinos, but this almost never happens as the proton has a half-life of 10^{31} years, a billion trillion times more than the age of the universe.

There's a little bit of Genghis Khan in all of us—quite literally, thanks to the indestructibility of atoms.

Around the world on a roll of toilet paper

▶ The average stream in the USA is contaminated with almost 1 milliliter of cholesterol in every million liters of water, a concentration equal to one sheet of toilet paper in a roll stretching two and a half times around the world.

A billion sheets of toilet paper, each ~10 cm (4 in) across, would stretch for 100,000 km (62,137 miles). To get this many sheets you would need to unravel:

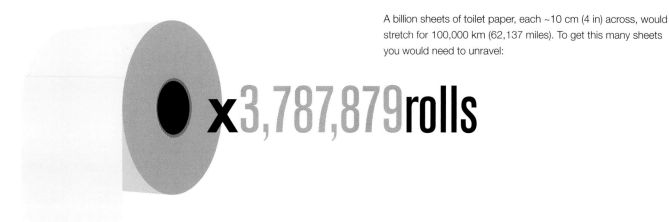

x3,787,879rolls

In chemistry, concentrations as small as one part in a billion are commonplace, especially when describing levels of water or airborne contamination with potentially dangerous chemicals. Such concentrations are also routinely met in the laboratory, in medicine, in industry, and elsewhere. But visualizing such tiny levels is difficult; recasting them at more familiar scales, or with more familiar objects than molecules and milliliters, can help.

An Olympic swimming pool has a volume of at least 2,500,000 liters.

If you had to fill one with a dropper that produced droplets of 6 mm³, it would take over:

x400,000,000,000

Atmospheric concentrations of carbon dioxide 1000–2004

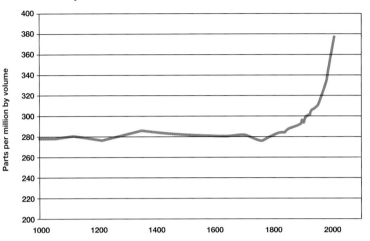

The level of carbon dioxide in the atmosphere has risen dramatically since the Industrial Revolution; even so, CO_2 makes up just a tiny proportion of the atmosphere. Preindustrial levels were 290 parts per million (ppm)—equivalent to about five hours in two years. Since then, concentrations have risen to 386 ppm.

A typical saltwater fish requires a dissolved oxygen level of at least **4 ppm**, equivalent to four large mouthfuls in a lifetime of eating.

In homeopathic medicines, the active ingredients are often diluted to concentrations lower than one part per trillion—equivalent to one mile on a two-month journey at the speed of light.

At concentrations of one part per quadrillion, the amount of active ingredient present in each drop of homeopathic medicine is equivalent to one human hair out of all the hair on all the heads of all the people in the world.

The dry-cleaning solvent perchloroethylene (PCE) is so **toxic** to humans that only very low concentrations are permitted by law: 5 parts per billion (ppb) in water. This is equivalent to five drops of gasoline in an entire railroad tanker car with a capacity of:

113,500 liters

Many everyday **pharmaceuticals** find their way into the water supply, but contamination levels are generally low. If a medicine is present at concentrations of 1 ppb, it will take someone drinking

3.8 liters/
1 gallon/
16 glasses

of water **every day** for years to consume the equivalent of one tablet of a common across-the-counter or prescription drug. Here are just three examples:

One tablet of Ritalin or Valium	**3.5 years**	
One capsule of Benadryl	**14.5 years**	
One tablet of Children's Tylenol	**58 years**	

The combined level of 33 suspected or known hormonally active compounds in U.S. streams is 57 ppb—equivalent to 57 kernels of corn in a 14 m (45 ft) high, 5 m (16 ft) diameter silo.

Australia covered in rice

▶ If every molecule in a quarter of a cup of table salt were a grain of rice, there would be enough to cover Australia to a depth of 1 kilometer (0.6 miles).

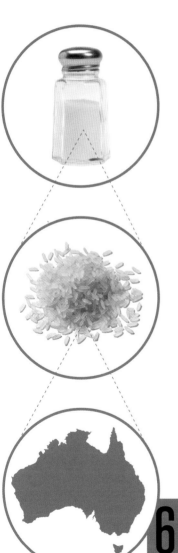

As we discovered on page 60, atoms are really tiny, and molecules made from atoms are not much bigger. This means that counting or weighing individual atoms or molecules of a substance is almost impossible—and certainly very impractical. Yet exact quantities are vital in chemistry, for instance making it possible to work out how many atoms of element X combine with how many atoms of element Y to give the compound XY. If you can tell, for example, that every atom of carbon used in an experiment has combined with two atoms of oxygen, you can conclude that the resulting compound has the formula CO_2.

So chemists are faced with a problem—how to measure out amounts of substances they can work with but still keep count of the number of particles. The solution is to use Avogadro's Number—a number that tells you how many atoms of an element are present in an amount that weighs the same in grams as the atomic mass of the element. For instance, the atomic mass of carbon is roughly 12, so 12 grams of carbon will contain Avogadro's Number of carbon atoms. This works for other types of particle as well. The molecular mass of water (the sum of the atomic mass of its component atoms) is almost exactly 18, so 18 g (or 18 ml) of water will contain Avogadro's Number of water molecules—this is known as a mole of water.

Avogadro's Number

6.0221367 x10²³ grains of rice

6.0221367 x10²³

Avogadro's Number is extremely huge—as many cupfuls of water as there are in the **Pacific**.

The most accurate laboratory balance can measure out **0.0001 g** of a substance, but this is still in the order of:

10¹⁹ atoms

One cubic centimeter of air—roughly the size of a sugar cube—contains:

45 million trillion molecules

It would take another 22,399 sugar cubes of air to give a mole—and therefore Avogadro's Number—of air molecules.

In a small block of **iron** the size of four sugar cubes, there are as many atoms as cents needed to make everyone in the world a trillionaire.

1,000,000,000,000,000

X

If each water molecule in a mole of water were an orange, there would be enough oranges to make a sphere the size of **Earth**.

If you had Avogadro's Number of pop cans, there would be enough to cover the entire **planet**.

Atomic bumper cars

▶ A dust particle suspended in water will appear to move around at random; the underlying process, known as "Brownian motion," is equivalent to a giant ball being battered on all sides by millions of tiny bumper cars.

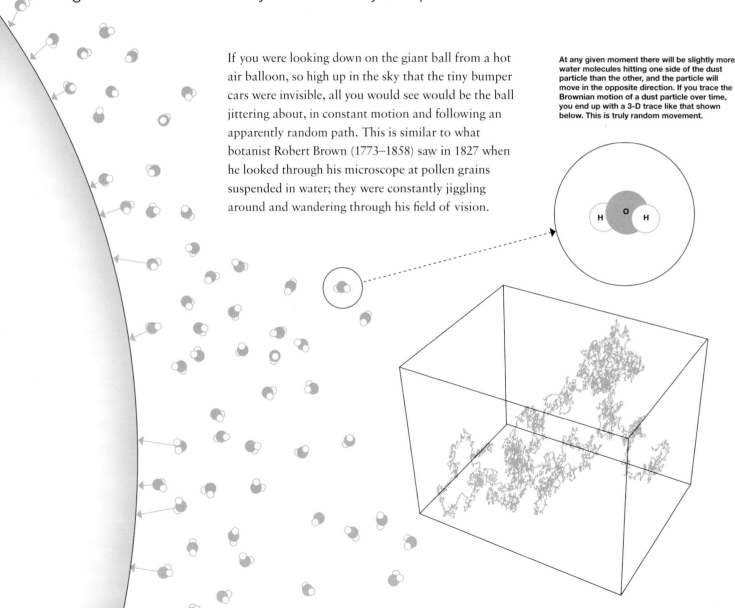

If you were looking down on the giant ball from a hot air balloon, so high up in the sky that the tiny bumper cars were invisible, all you would see would be the ball jittering about, in constant motion and following an apparently random path. This is similar to what botanist Robert Brown (1773–1858) saw in 1827 when he looked through his microscope at pollen grains suspended in water; they were constantly jiggling around and wandering through his field of vision.

At any given moment there will be slightly more water molecules hitting one side of the dust particle than the other, and the particle will move in the opposite direction. If you trace the Brownian motion of a dust particle over time, you end up with a 3-D trace like that shown below. This is truly random movement.

H   O   H

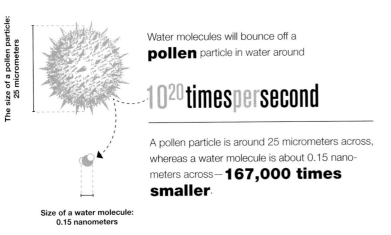

The size of a pollen particle: 25 micrometers

Water molecules will bounce off a **pollen** particle in water around

10^{20} **times per second**

A pollen particle is around 25 micrometers across, whereas a water molecule is about 0.15 nanometers across—**167,000 times smaller**.

Size of a water molecule: 0.15 nanometers

Brownian motion of the pollen particle is equivalent to a balloon **10 meters** (32 ft) across being batted around by the crowd in a football stadium entirely filled with fans.

650

Average speed of a water molecule at room temperature is

meters per second

If unimpeded, a water molecule could travel 4 trillion times its own width, but in fact the molecule will change direction at least 100 billion times a second due to collisions with its neighbors.

The mathematics of Brownian motion can be used to model many other systems: the stock market, for instance. The price of a stock jitters about as millions of investors buy and sell shares, just as a pollen grain jitters about in **Brownian motion**.

Having ruled out water currents, he repeated his observations with long-dead pollen and then with rock dust—any particles small enough showed the same movement. This phenomenon later came to be named Brownian motion.

At the time, the cause of the motion was unknown, but 50 years later it was suggested that it could be due to the impact of moving water molecules. Up until the beginning of the 20th century, however, the atomic theory of matter was still regarded as unproven speculation; many people doubted that such things as atoms and molecules really existed. Einstein's 1906 paper on Brownian motion, which gave testable predictions, led to proof that the motion was indeed due to the bombardment of tiny, atomic-scale particles. In fact, measurements of Brownian motion could be used to work out the size of atoms and by extension Avogadro's Number (see the previous entry).

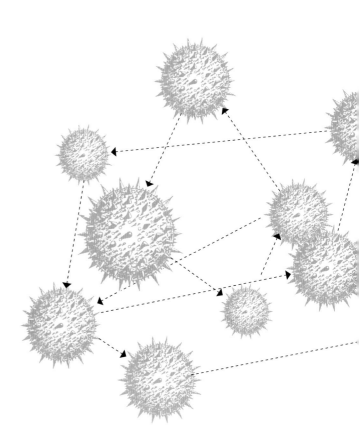

Far away so close

▶ From a distance, a beach looks solid, but close up it is possible to see the tiny grains of sand; similarly, only under extreme magnification is it possible to see that a solid is composed of discrete particles.

Zoom in on a single square centimeter patch of beach and you'll be looking at around

2,500 grains of sand.

Looking at a solid it is hard to accept that it is made up of tiny particles. At a human—or macroscopic—scale, solid surfaces look smooth and continuous, and it is counterintuitive to think of them being made up of discrete particles. As with the sandy beach, however, this is a matter of perspective and scale. Get close enough to the beach and you can see the individual grains of sand, although a better analogy might be with wet sand, in which the particles are bound together; dry sand is arguably more analogous with a liquid. An alternative analogy is with a photograph in a newspaper. From a distance, the shapes and blocks of shade look continuous, but up close they prove to be composed of dots that do not touch.

Sand is a very unusual substance; when dry it acts more like a **liquid** (it has a fixed volume but not shape—in other words, it can be poured) and when wet it acts more like a **solid**.

What's the difference between a bucket of sand and a bucket-shaped lump of solid quartz? Both contain almost identical amounts of silicon dioxide, but only one of them can be poured through a sieve—because in solid quartz every molecule of silicon dioxide is bonded to another one.

1,000,000,000,000

It is estimated that the total number of **grains of sand** on Earth may be as high as:

$$10^{24}$$ **(a trillion trillion)**

Sand is made of silicon dioxide, the most abundant compound in Earth's crust, making up 42.86% of the crust by weight.

42.86%

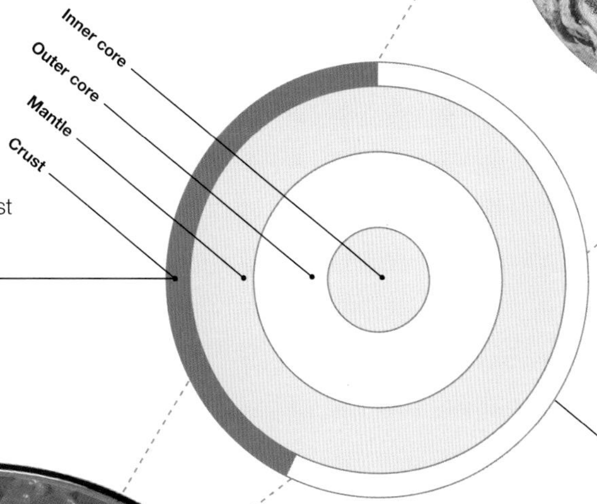

Inner core
Outer core
Mantle
Crust

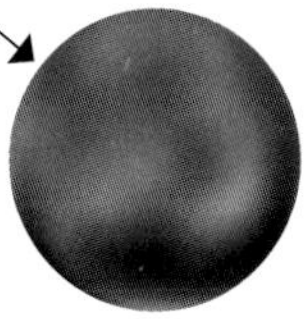

This would be enough to make a planet the size of **Pluto**.

If all of the **silicon dioxide** in the crust were ground into sand, there would be a pile of sand weighing

$$1.187 \times 10^{22} \;\; \text{kg}$$

X
1,000,000,000,000,000
1,000,000,000

Everybody dance now!

▶ A solid changing phase into liquid and then into a gas is like a bunch of dancers packed onto a dancefloor, who start off unenthusiastically and then become progressively more excited until they are leaping about.

According to the kinetic theory of matter (see pp. 38–9), atoms and molecules all have some degree of motion, even when chilled to within a whisker of absolute zero. What distinguishes the three main phases of matter—solid, liquid, and gas—is the amount of motion. In a solid, the atoms (and molecules) are like dancers, packed onto a dancefloor so tightly they are forced to link arms, who don't particularly like the music that is playing, and so are merely shifting from foot to foot without enthusiasm.

Increasing the temperature so that the solid melts to give a liquid is like putting on a track that the dancers really like, so that although still tightly packed and confined to the dancefloor, they begin to dance properly, moving about and changing partners. Increasing the temperature further so that the liquid boils into a gas is like cranking up the volume and sending the dancers into a frenzy, so that they begin to leap and cavort, spreading out to take up the entire nightclub and only touching when they bash into one another in the course of their wild gyrations.

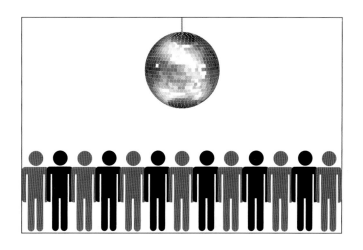

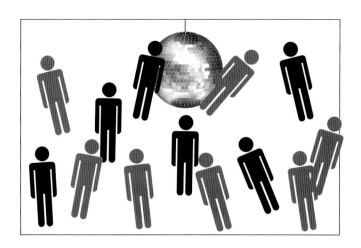

Zero on the Kelvin temperature scale—known as **absolute zero**—is a state where matter has no energy whatsoever, and is impossible to reach. The lowest temperature ever achieved, probably in the history of the universe, is 100 picoKelvin, or 0.1 billionth of a Kelvin (0.0000000001 K), by the Low Temperature Laboratory at the Helsinki University of Technology.

61

At temperatures close to absolute zero, the speed of light can be reduced to

kilometers per hour (38mph)

Graphite, a form of carbon, has a melting point of:

3,500°C

Diamond, another form of carbon, has a boiling point of

4,027°C

In a gas, each molecule or atom is on average ten times its own diameter away from the next one. When you squeeze a balloon and it resists the pressure of your hands, you are squeezing 90% empty space—the resisting force of the gas inside comes from the extremely high velocity of the gas particles.

At room temperature the average speed of an atom is

~500 m/s (~1,100mph)

At 200 microKelvin, atoms move at

20 cm/s (~0.45mph)

Pluto:
Average surface temperature

40K (−233°C / −388°F)

The universe:
Average background temperature

2.72778K (−270.4222°C / −454°F)

Helium:
Boiling point

4.222K (−268.9278°C / −452°F)

He

Space:
Coldest point (Boomerang Nebula)

1K (−272.15°C / −457°F)

Boy meets girl

▶ A chemical reaction reaching equilibrium is like a school dance, with as many students hitting the dancefloor as getting bored and sitting down.

The school dance is being held in the gymnasium. At the start of the evening, the students come in and stand around the dancefloor. The bravest boy asks a girl to dance, soon followed by others. Eventually each couple gets bored of dancing together, so they separate and go back to the edge of the dancefloor, but to start with there are far more unpaired students eager to find a dance partner than there are dancing couples, so the dancefloor quickly fills up. Eventually, however, there are so many dancing couples that it becomes hard for unpaired students to find a partner, so the rate at which new couples go onto the dancefloor slows down. Meanwhile, more dancing couples means more students getting bored and leaving the dancefloor. In time the rate at which students are joining the dance-floor will equal the rate at which they are leaving it, and the system is said to be in dynamic equilibrium.

Boy Girl

This is analogous to what happens when a reversible reaction reaches a dynamic equilibrium. Chemicals A and B begin reacting together to produce AB, but the reaction is reversible and AB can also split apart. To start with, there is lots of A and lots of B, but not much AB, so there's a lot more coupling up than splitting apart. But eventually AB is splitting apart as quickly as A and B are coming together, and the reaction is said to have reached equilibrium.

The school dance analogy makes it possible to work out how to shift the equilibrium to produce more dancing couples (i.e., more AB). One way is to increase the number of students, or to make the gymnasium smaller so that single students can find partners more quickly: these are equivalent to boosting the concentration. Boosting the tempo of the music encourages single students to look for dance partners: this is like increasing the temperature.

When students are separating and sitting down at the same rate as they are getting up to dance with each other, the school dance has reached equilibrium.

"equilibrium"

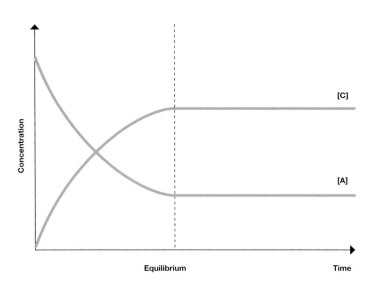

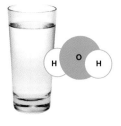

If A is the concentration of reactants in a solution, and C is the concentration of products, this graph shows how concentrations change with a simple equilibrium reaction where A $\rightleftharpoons$ C.

The graph below shows how the rate of the forward reaction slows and the rate of the reverse reaction increases, until the two are equal, at which point the system is in a state of dynamic equilibrium.

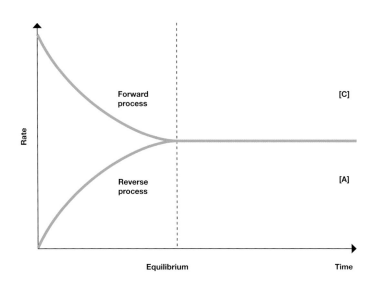

An important equilibrium reaction in industry is the reaction that produces **ammonia**, used in vast quantities for the manufacture of fertilizer. In 2007, world production of ammonia was

131 **million tonnes**

It is estimated that by **2013**, ammonia production worldwide is likely to reach

218 **million tonnes**

Roughly equivalent to the amount of **garbage** produced each year in the USA.

In liquid water, water molecules form, break, and reform H-bonds with other molecules in a dynamic equilibrium. Every molecule in a glass of water is changing partners **billions** of times a second.

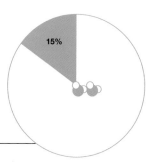

As surprising as it may seem, at any one moment, only 15% of the molecules in a glass of water are actually **touching**.

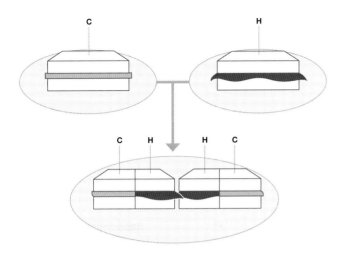

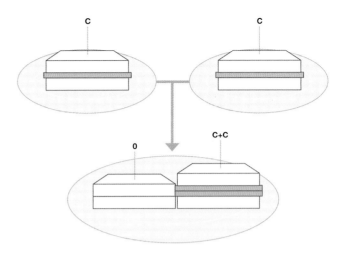

Covalent bond

Two atoms form a covalent bond when they share each other's valence (bond-forming) electron, like two people sharing their sandwiches.

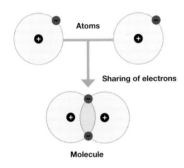

Ionic bond

In an ionic bond there is no real sharing—one atom simply takes the electron of the other one, leaving the donor positively charged, while the receptor becomes a negatively charged ion.

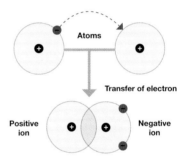

Shared sandwiches and chemical bonds

▶ The different types of chemical bond are like different ways of sharing sandwiches; for instance, a covalent bond is like sharing half of each other's sandwich.

Chemical bonds between atoms are formed through the interaction of their electrons. With ionic bonds, one atom donates an electron, and the other atom receives it. With covalent bonds, two atoms are said to "share" electrons, but it is not clear what this means. One way of looking at it is through a sandwich analogy. If you give me half of your cheese sandwich and I give you half of my ham sandwich, we still have our original sandwiches but we also have new flavors—we have the snack equivalent of a covalent bond.

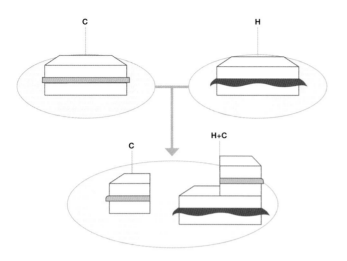

Polar covalent bond

Although the two atoms share an electron orbital, creating a covalent bond, the orbital is unequally distributed, so that one atom gets "more" of the electron, becoming partially negatively charged.

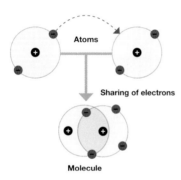

Other types of bond correspond to other ways of sharing sandwiches. Unequal sharing of sandwiches, so that I give you half of my sandwich but you only give me a quarter of yours, is like a polar covalent bond. If I steal your sandwich and don't give you any of mine, I am like an atom accepting two donor electrons in a dative or dipolar bond.

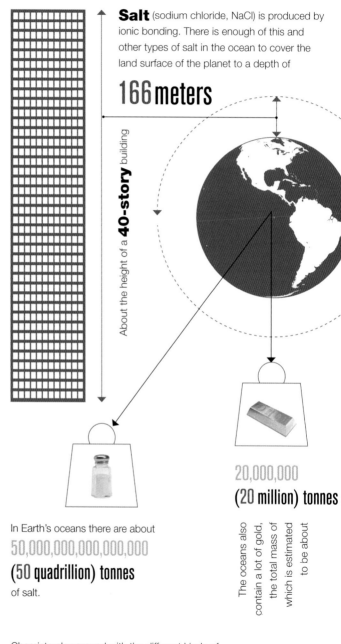

Salt (sodium chloride, NaCl) is produced by ionic bonding. There is enough of this and other types of salt in the ocean to cover the land surface of the planet to a depth of

166 meters

About the height of a **40-story** building

In Earth's oceans there are about
50,000,000,000,000,000
(50 quadrillion) tonnes
of salt.

The oceans also contain a lot of gold, the total mass of which is estimated to be about

20,000,000
(20 million) tonnes

Chemists play around with the different kinds of bonding in a process known as combinatorial chemistry, which allows them to synthesize multiple different compounds at a time, up to:

40,000

All tied up

▶ A water drop forming a sphere because of surface tension is like a crowd in which each person is tied to their nearest neighbors.

Imagine a crowd of people in a field, where each person is tied to all the people surrounding them. The people are restless and tend to pull on the ropes binding them together, with the result that a woman in the middle of the crowd feels more or less the same pull from all sides. The people on the edge of the crowd, however, don't have anyone in front of them to be tied to; they are only tied to the people to the left and right and behind them. As a result, they are not pulled equally on each side, and the inevitable consequence is that the crowd will form into a circle. The crowd of people is experiencing surface tension.

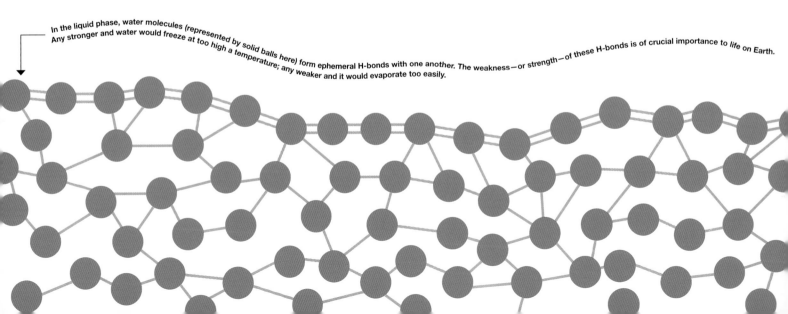

In the liquid phase, water molecules (represented by solid balls here) form ephemeral H-bonds with one another. The weakness—or strength—of these H-bonds is of crucial importance to life on Earth. Any stronger and water would freeze at too high a temperature; any weaker and it would evaporate too easily.

The same thing happens in a crowd of water molecules. Molecules of water form hydrogen bonds (weak and relatively easily broken electrostatic bonds) with one another, and these H-bonds are responsible for many of the remarkable and unique properties of water, including surface tension. Like the people on the edge of the crowd, water molecules at the surface of a mass of water are pulled only to the sides and backward, making it hard to displace them away from the bulk of the water molecules, and causing any free mass of water to adopt the shape with the least surface area—a sphere.

An average ice cube contains about **6×10^{22} molecules**. If you had come up with a different arrangement of these molecules every week since the Big Bang, you would still not have exhausted the number of possibilities. This is why every ice cube ever created will probably have had a different arrangement of water molecules.

A liquid will wet a surface (as opposed to sitting on it as droplets) if the angle at which it makes contact with the surface is more than.

90°

If H-bonding was not present in water, it would boil at –90°C, and there would be almost **no liquid** water on Earth.

–90°C

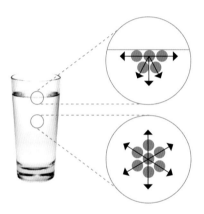

Below the surface, water molecules pull each other equally in all directions. At the surface they pull each other downward and sideways, creating surface tension.

4°C

This is the temperature at which fresh water reaches maximum density. Because of this, it will expand whether heated or cooled. Water is the only substance on Earth that doesn't achieve maximum density when solidified.

When water freezes each molecule forms four relatively rigid H-bonds with its neighbors. This acts to push the molecules further apart than during the liquid phase, and as a result ice is less dense than water.

Ham sandwiches and stoichiometry

▶ Producing a chemical reaction is like making a sandwich: to make a ham sandwich you need to add the right amount of ham to the right amount of bread. Likewise, to make carbon dioxide, you need to react the right amount of oxygen with the right amount of carbon.

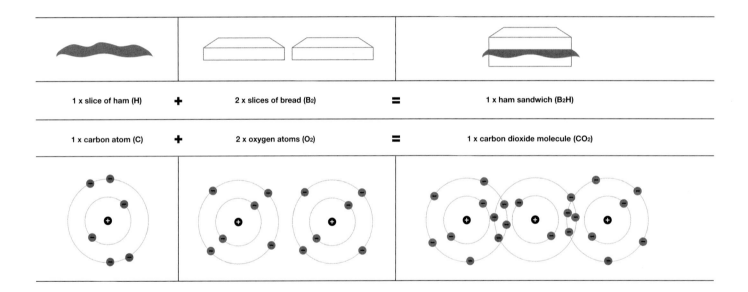

| 1 x slice of ham (H) | **+** | 2 x slices of bread (B₂) | **=** | 1 x ham sandwich (B₂H) |

| 1 x carbon atom (C) | **+** | 2 x oxygen atoms (O₂) | **=** | 1 x carbon dioxide molecule (CO₂) |

The branch of chemistry that deals with the arithmetic of reactions—specifically the precise quantities of substances that are reacted and produced in chemical reactions—is known as stoichiometry. An example of stoichiometry is working out the formula of a compound from the proportions of the mass of each reactant and product. The combination of chemical equations, math, and a fearsome-sounding name make stoichiometry the scourge of the chemistry classroom, but in fact it can be child's play, as the analogy with sandwich-making demonstrates.

Imagine that you are making ham sandwiches, and you know that a slice of bread (B) weighs 10g and a slice of ham (H) weighs 5g. If you have 125g of ham, how much bread will you need to turn all the ham into ham sandwiches? The recipe (or formula) for a ham sandwich is simple: two slices of bread for every slice of ham, or B₂H. Thus all you need to do is calculate how many slices of ham you have, multiply this number by two to give the number of slices of bread you will need, and then multiply

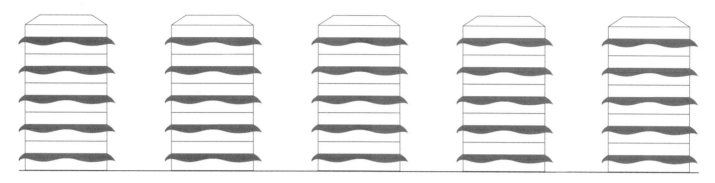

You don't have to count them to know there are 25 ham sandwiches here—you could just weigh them. As long as you know how much slices of ham and bread weigh, and the formula for a ham sandwich, you will be able to calculate the number of sandwiches present.

The Chemical Abstracts Society (CAS) assigns ID numbers to every chemical that is discovered or synthesized. Currently, the CAS lists more than **132 million** unique organic and inorganic substances; more than are added

12,000 each day

this number by the weight of a single slice of bread: [(125/5) x 2] x 10 = 500 g of bread. How does this apply to chemistry? Just replace ham with carbon and bread with oxygen, and using the same simple formula you can just as easily work out the mass of oxygen required to completely combust 125 g of carbon in the reaction $C + O_2 = CO_2$. (125/12 [the atomic mass of C] x 2) x 16 [the atomic mass of O] = 333 g of oxygen.

The **number of possible chemicals** depends on the number of elements and their compounds, and the number of ways these can be put together to form new compounds. Estimates of the total number of possible chemicals range from 10^{18} (close to the number of grains of sand on Earth) to 10^{200} (at least 10^{100} more than the number of particles in the universe). If the latter estimate is true, then obviously not every possible combination physically exists in the universe.

One man who could have benefited from stoichiometry was 17th-century alchemist **Hennig Brand**, who isolated pure phosphorus by putrefying urine in his basement. He needed **60 barrels** of urine to isolate a tiny quantity of **phosphorus**.

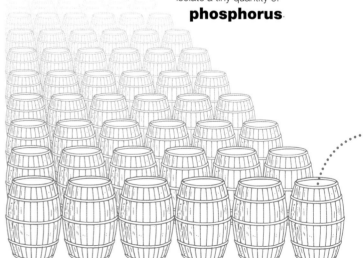

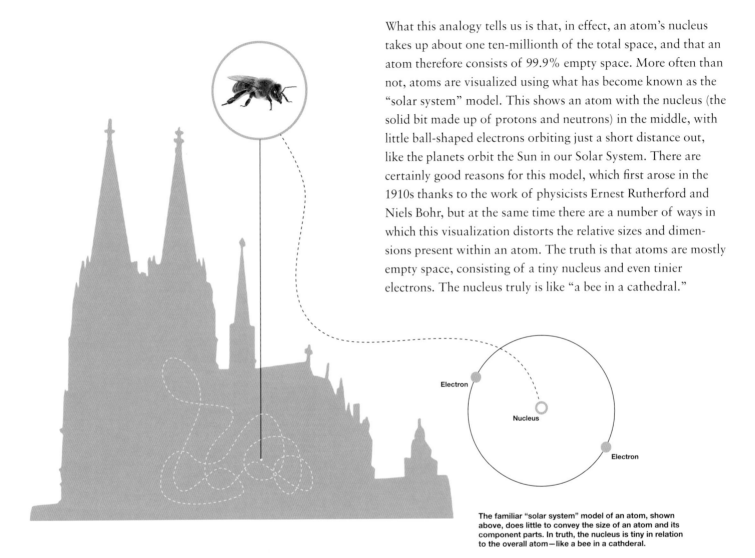

What this analogy tells us is that, in effect, an atom's nucleus takes up about one ten-millionth of the total space, and that an atom therefore consists of 99.9% empty space. More often than not, atoms are visualized using what has become known as the "solar system" model. This shows an atom with the nucleus (the solid bit made up of protons and neutrons) in the middle, with little ball-shaped electrons orbiting just a short distance out, like the planets orbit the Sun in our Solar System. There are certainly good reasons for this model, which first arose in the 1910s thanks to the work of physicists Ernest Rutherford and Niels Bohr, but at the same time there are a number of ways in which this visualization distorts the relative sizes and dimensions present within an atom. The truth is that atoms are mostly empty space, consisting of a tiny nucleus and even tinier electrons. The nucleus truly is like "a bee in a cathedral."

Electron

Nucleus

Electron

The familiar "solar system" model of an atom, shown above, does little to convey the size of an atom and its component parts. In truth, the nucleus is tiny in relation to the overall atom—like a bee in a cathedral.

A bee in a cathedral

▶ If an atom were blown up to the size of a cathedral, the nucleus would be no larger than a bee buzzing about in the center, while the electrons would be "orbiting" near the outermost edges.

Even if the atom were blown up to the size of the largest sports stadium, electrons would remain far too small to be seen. In fact, it is thought that electrons have no dimensions at all, but are mere "point-particles"—dimensionless points of mass in space.

The nucleus, though small, does have dimensions, as do the protons and neutrons from which it is made. But the tiny particles that make up protons and neutrons—quarks—do not. They, like electrons, are merely points of mass and energy, with no physical dimensions.

Imagine the average atomic nucleus expanded to the size of a marble. At this scale, it would weigh something in the order of . . .

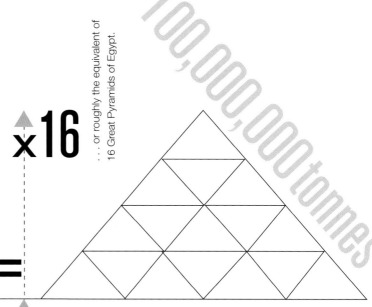

... or roughly the equivalent of 16 Great Pyramids of Egypt.

×16

100,000,000 tonnes

100,000,000 tonnes

Protons are so small that 500 billion of them can fit on the **head of a pin**. The latest research suggests that a proton is about 1.6×10^{-15} m in diameter.

500 billion

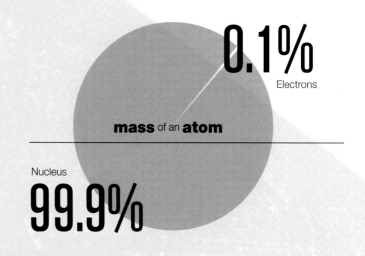

0.1%
Electrons

mass of an **atom**

Nucleus
99.9%

Despite being small, the nucleus is heavy—99.9% of the atom's mass lies within the nucleus, leaving only 0.1% for the electrons.

Solar systems and quantum fans

▶ The atom is like the optical illusion created by an electric fan, with the electrons occupying a "cloud" of potential locations around the nucleus at the hub.

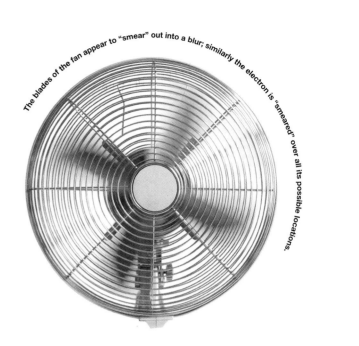

The blades of the fan appear to "smear" out into a blur; similarly the electron is "smeared" over all its possible locations.

In 1911 the physicist Ernest Rutherford (1871–1937) came up with a startling new theory about the atom to explain some unexpected results. Atoms had previously been believed to be solid, cubelike structures, or perhaps like plum puddings, with negative particles (electrons) studded around in a "soup" of positive matter. But an experiment in which subatomic particles had been fired at a thin film of gold atoms had shown that while many of them passed straight through, some rebounded as if they had impacted a dense concentration of mass. This suggested to Rutherford that the atom must be mostly empty space, with the vast majority of its mass concentrated in a central nucleus that was minuscule compared to the overall size of the atom. In 1913 the Danish physicist Niels Bohr (1885–1962) developed Rutherford's model to present a picture of the atom as a miniature solar system, with the nucleus as the Sun and the electrons orbiting at fixed distances like the planets in their orbits, and with electrostatic attraction in place of gravity. This model is still useful, as it demonstrates how the electrons jump between discrete energy levels as they capture or lose packets

The "Electron Cloud" Model

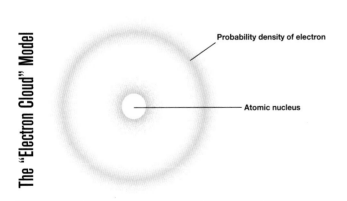

Probability density of electron

Atomic nucleus

The "Solar System" Model

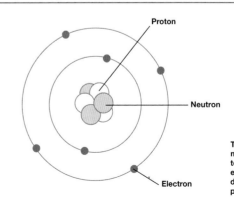

Proton

Neutron

Electron

The "solar system" atomic model was a significant step toward today's model, with electrons orbiting a hard, dense nucleus made of both protons and neutrons.

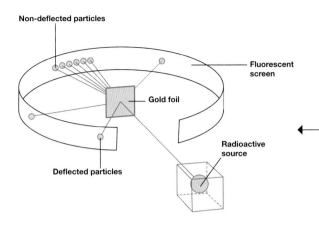

Non-deflected particles

Fluorescent
screen

Gold foil

Radioactive
source

Deflected particles

The **gold foil experiment** enabled Rutherford to discard the pudding model of the atom, which predicted that particles fired at a heavy atom would become embedded in the pudding. In fact, particles fired at the gold foil mostly passed right through, with a few rebounding violently.

99.999999999999%

of an atom's volume is just empty space!

(quanta) of energy. Later quantum mechanics proved that it is not possible to determine with certainty the location of an electron, only to describe its probable location, so instead of an orbit an electron is now said to occupy a cloud of possible positions, and to be in some senses in all of those positions at once. The electron is in effect smeared across the cloud of possible locations, with "more" in the areas where the probability of its location is highest. Visually, this model resembles what you see when you look at an electric fan; the moving blades blur into a disc that seems both solid and insubstantial.

If the protons and neutrons that make up the nucleus were each a centimeter wide, then electrons and quarks would be less than the diameter of a strand of hair and the entire atom's diameter would be greater than the length of 30 football fields.

30x

The **"Plum Pudding" Model**

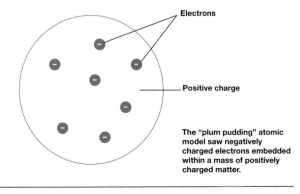

Electrons

Positive charge

The "plum pudding" atomic model saw negatively charged electrons embedded within a mass of positively charged matter.

The **ratio** of the diameter of the nucleus: atomic diameter is

1:100,000

If all the empty space in the atom could be removed, the **entire human race** could be packed into the volume of a sugar cube.

Extremely bright
and incredibly dim

▶ Using a logarithmic scale is like using a filter to equalize light sources of very different brightness, making it possible to register ones that would otherwise be drowned out.

Imagine you have a dozen different lights aimed at you, and you want to pinpoint the source of each light. If one of the lights is blindingly bright, while the others are all tiny pinpricks, it will be impossible to see any but the bright source. If you use a special filter that, say, translates different brightness levels into different colors of the same brightness, then you could locate all the light sources and determine their brightness.

A logarithmic scale is the mathematical equivalent of this filter. If you need to plot a graph using numbers that differ by several orders of magnitude—for instance when discussing acidity (i.e., concentration of the acidic hydronium ion, H_3O^+)—you will either need to use a piece of graph paper the size of a country, or squish all the smallest data points down one end of the graph. However, if you take the logarithm of each number (essentially the power to which it is raised), you get a set of data that is much easier to compare. For example, the hydronium concentration of household ammonia (10^{-11}) is 10 billion times smaller than that of automobile battery acid (roughly 10^{-1}). Plotting this on a normal graph with any intervening data points would be impossible, but plotting the logarithms (−11 and −1) is simple. Also a log axis records percentage change, rather than absolute change—e.g., 10 and 12 will be equally far apart as 50 and 60 on a log axis because in each case there is a 20% increase.

X 10 million is the number of pages you'd have to extend the height of this book by to accommodate the Sun's brightness on this linear chart.

4,000,000,000 candelas (the Sun)

120 candelas (100-watt light bulb)

1 candela (a candle)

Linear scale

Logarithmic scale

A log scale preserves absolute quantitative information but also shows the relationships between data points, making it much easier to read qualitative information from a graph or chart.

4,000,000,000	4,000,000,000 candelas
400,000,000	
40,000,000	
4,000,000	
400,000	
40,000	
4,000	
400	120 candelas
40	
4	
0	1 candela

Other logarithmic scales include the **Richter** and moment magnitude scales for measuring earthquakes, and the decibel (dB) for measuring power, especially acoustic power.

The pH scale is logarithmic; as a result, each whole pH value below 7 is ten times more acidic than the next higher value. For example, pH 4 is ten times more acidic than pH 5 and 100 times more acidic than pH 6.

Concentration of hydrogen ions compared to distilled water	pH number	Examples of solutions and their pH strength
1/10,000,000	14	Drain cleaner; caustic soda
1/1,000,000	13	Bleach; oven cleaner
1/100,000	12	Soapy water
1/10,000	11	Household ammonia
1/1,000	10	Milk of magnesia
1/100	9	Toothpaste
1/10	8	Eggs; seawater
0	7	Purified water
10	6	Milk
100	5	Coffee
1,000	4	Tomato juice
10,000	3	Orange juice
100,000	2	Vinegar
1,000,000	1	Hydrochloric acid
10,000,000	0	Sulfuric acid

Increasingly alkaline ▲

Increasingly acidic ▼

Molecular zippers

▶ DNA and RNA are made up of complementary pairs of long-chain molecules that can zip up and unzip like the zipper on your jacket.

When Francis Crick and James Watson worked out the structure of DNA in 1953, they immediately noticed that the unusual and elegant form they had elucidated went hand in hand with the mechanism by which DNA carried out its function—the ability to store and replicate genetic information. X-ray crystallography studies had shown that DNA is a long chainlike molecule, constructed like a ladder that is twisted into a helix. The "rungs" of the ladder are composed of four varieties (A, G, C, and T) of the chemical group known as bases, arranged in pairs. The crucial piece of the puzzle fell into place for Crick and Watson when they learned that the base pairs only occurred in a limited number of configurations: A with T, and C with G.

It was obvious that this was the key to DNA's ability to replicate itself, and Crick and Watson went straight to the pub to declare they had discovered the secret of life. What they had realized was that if DNA comes apart down the center, like a zipper being opened, the base pairs act like the teeth of the zipper. As the zipper opens, the interlocking "teeth" are split apart, and the now "naked" row of teeth on each side can serve as the template for new matching sequences to be assembled. It is now known that DNA and its close chemical cousin RNA do indeed work like zippers, with a special enzyme known as helicase acting like the zipper slider.

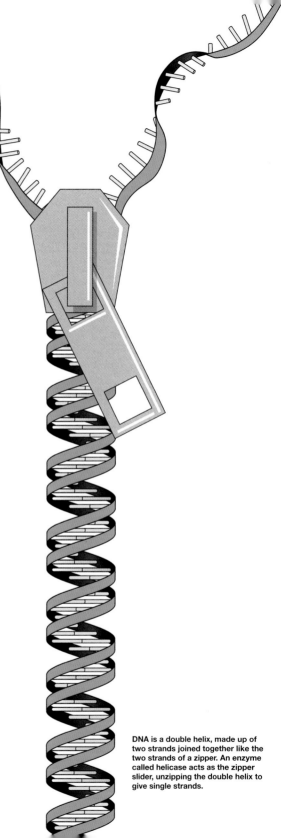

DNA is a double helix, made up of two strands joined together like the two strands of a zipper. An enzyme called helicase acts as the zipper slider, unzipping the double helix to give single strands.

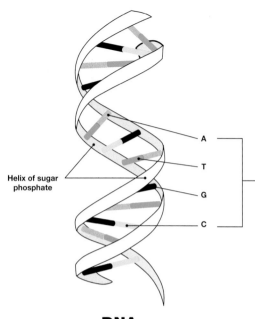

DNA

Deoxyribonucleic acid

Helix of sugar phosphate

A
T
G
C

Adenine — Thymine

Guanine — Cytosine

If all **3 billion** "letters" in the human genome were stacked 1 millimeter apart, they would form a pile 7,000 times higher than the Empire State Building.

The modern zipper was invented by **Gideon Sundbach** in 1917. Today most of the world's zippers are made by YKK, which stands for Yoshida Kogyo Kabushikikaisha.

Today, YKK have **206 factories** in 52 countries, making everything from the brass for the zippers to the dyed fabric that surrounds them, and they produce **2,000 kilometers** (1,200 miles) of zipper every day in their Macon, Georgia factory alone—that's enough zipper to reach from Berlin to Rome. In a total of **1,500 styles** in more than 427 standard colors, the Georgia factory produces

7 million

zippers a day
(over 2 billion a year!)

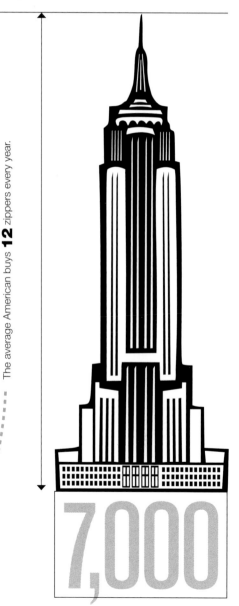

The average American buys **12** zippers every year.

7,000

Macromolecules and mega moles

▶ If a water molecule were blown up to the size of a dime, a nucleic acid molecule would be 10 centimeters (4 inches) thick and up to several hundred kilometers long.

There is enormous variation in the size of molecules, especially in the world of organic chemistry (the chemistry of carbon). Carbon is able to bond to other atoms of carbon to produce long chains—theoretically infinitely long. This means that the molecules encountered in nature are often massive compared to the ones in school laboratories. For instance, while table salt (NaCl) has a molecular weight of 58.433 g/mol, a complex organic molecule such as chlorophyll A ($C_{55}H_{72}O_5N_4Mg$) has a molecular weight of 893.51 g/mol.

With a long, thin rounded shape like a baguette, a nucleic acid molecule enlarged 5 million times would remain relatively narrow while stretching to an incredible length.

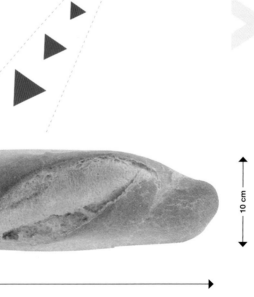

×5,000,000

10 cm

several hundred kilometers

I——— **1.35 mm thick**

US dime

17.91 mm diameter

Water is composed of simple molecules, each containing just three atoms, so they are comparative tiddlers in the molecular world; macromolecules are the true behemoths of this microscopic realm.

Molecules with molecular weight higher than 10,000 are known as **macromolecules**. Macromolecules exist with molecular weights as high as 10 billion.

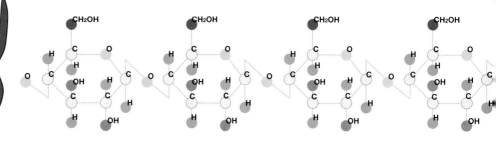

Cellulose is the most abundant organic compound on Earth; it has a molecular weight of at least 570,000.

DNA is one of the largest macromolecules. The DNA of the common bacterium ***E. coli*** contains about 3 million base pairs: its molecular weight would be around 1.8 billion g/mol.

Even the biggest molecules are **microscopic** on a human scale. Despite its vast molecular weight, a strand of DNA is so tiny, you could fit about

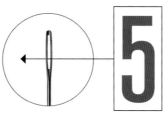

5 million of them through the eye of a needle.

If all the DNA in a human body was joined together to give one strand, it would be over **300 billion km** (186 billion miles) long; enough to stretch to the Moon and back 390,000 times, or to the Sun and back 1,000 times.

Balls and springs

▶ A molecule is like a series of balls connected together by springs.
The balls are like atoms and the springs are like chemical bonds.

When scientists like Crick and Watson (see pp. 90–1) or the Nobel prize-winning American chemist Linus Pauling were trying to work out the structure of molecules such as DNA, they made models out of sticks and balls. This stick-and-ball model has been updated to the spring-and-ball (SB) model; the springs represent the modern understanding of interatomic bonds as flexible, energetic, and in constant motion. But the idea that molecules in nature really are like microscopic arrangements of sticks and balls dates back to at least the mid-19th century, and has been so influential that even the basic belief that molecules have characteristic structures and shapes originates with the analogy.

So is it true? Something called the Born-Oppenheimer approximation—a mathematical system for computing the quantum mechanics of molecules—suggests that the SB model is a good description of how molecules really work. According to Professor Giuseppe Del Re, it is "as if the SB model a chemist has on his desk were a faithful magnification (roughly a hundred million times) of a molecule."

Modeling macromolecules (see pp. 92–3) has always been a challenge. One of the first-ever efforts—a model of myoglobin created by John Kendrew in 1958—involved a cube 2 m (6 ft) on a side, packed with a forest of 2,500 brass rods.

Graphite

Making a ball-and-spring model of a **DNA strand** is hard work. Modeling a stretch of DNA just 200 bases long would require around

6,000balls and at least that many springs.

Modeling using SB-like analogies is still at the cutting edge of science. Possibly the hardest substance ever discovered is ultra-hard high-pressure **graphite**—when squeezed between two diamond jaws at pressures of 170,000 atmospheres, this substance can crack the diamond. Modeling is revealing the secrets of its structure, raising the prospect of being able to manufacture super-hard materials.

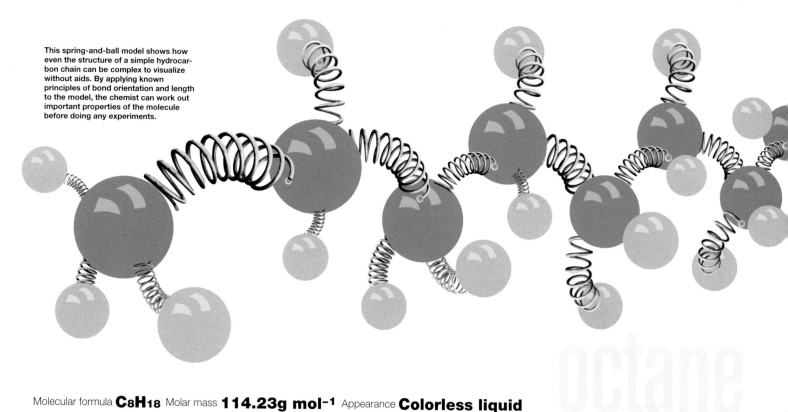

This spring-and-ball model shows how even the structure of a simple hydrocarbon chain can be complex to visualize without aids. By applying known principles of bond orientation and length to the model, the chemist can work out important properties of the molecule before doing any experiments.

Molecular formula **C_8H_{18}** Molar mass **114.23g mol^{-1}** Appearance **Colorless liquid**

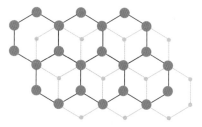

A model of the hexagonal arrangement of carbon atoms in graphene.

SB modeling suggests that in **graphene** a flat layer of carbon atoms is packed into a hexagonal arrangement, which gives it incredible properties. Being ~1 atom thick, it is transparent, yet a sheet stretched over a cup could support the weight of a truck (c.3,000 kg, or 6,600 lbs), even if all the weight were concentrated in a point the diameter of the tip of a pencil.

Knocking is a problem in combustion engines when air and fuel combust out of sync with the cylinders; octane is one of the most knocking-resistant types of fuel, so anti-knocking capacity is measured in comparison with octane to give fuel an octane rating. While the highest octane rating that octane can have is 100, ethanol has an octane rating of 129.

129

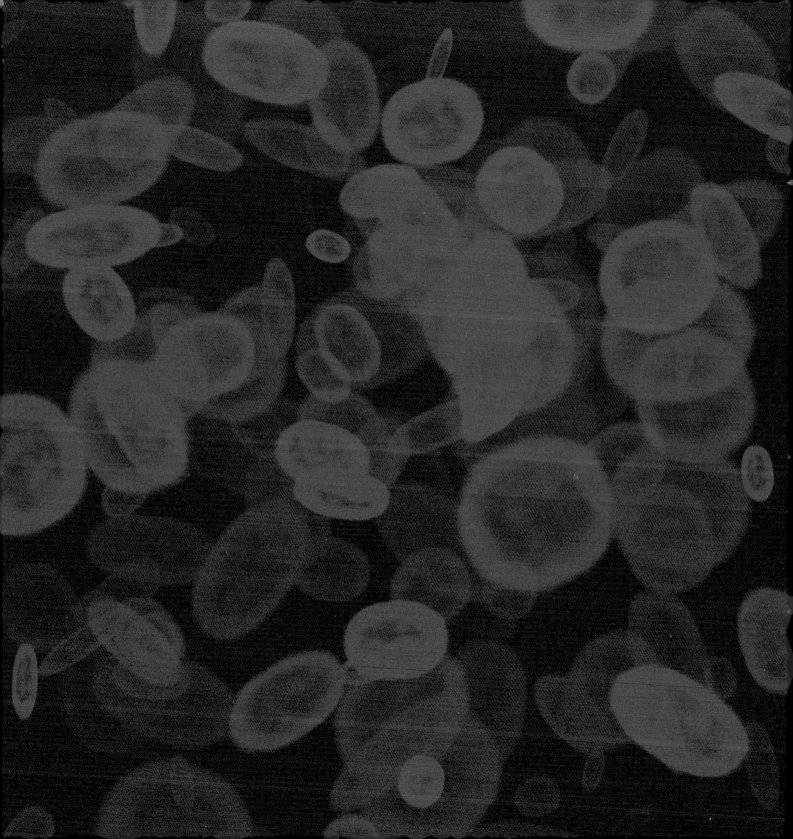

Section Three

▶ The realm of living things on Earth is massive and diverse, and as such it can be challenging to fully understand. This section presents a number of key biological principles. Whether it is comparing the living cell to manmade machines or comparing the speed of animals to that of vehicles, an analogy can serve to bring the biological world into perspective.

Biology

The history of Earth in a day

▶ If the entire 4.6 billion-year history of Earth were compressed into a day, the dinosaurs would have appeared at 22:46 and the first cities just a tenth of a second before midnight.

Visualizing the length of human history is hard enough, but when put in the context of the history of the planet it is almost impossible to square historical timescales with geological ones. Even incredibly slow processes, such as continental drift, flash past in the blink of an eye as far as the whole history of planet Earth is concerned.

One way round this is to use a logarithmic timescale (see pp. 88–9). Another is to reframe the history of Earth on a more familiar scale, such as the length of a day. An hour of this "cosmic day" is equivalent to roughly 187.5 million years, a minute to 3.125 million years, and a second to 52,000 years. If Earth were formed at midnight, the first life would appear surprisingly early—at 04:10. It would be another nine hours until the first complex cells appeared—up until 13:02 the planet would be the exclusive domain of the single-celled prokaryotes, the group that includes bacteria. By this time, however, photosynthesizing bacteria would already have pumped vast quantities of oxygen into the atmosphere.

The first animals would only have appeared at 18:47, and around two hours later episodes of dramatic cooling would have seen the planet frozen to give periods of "Snowball Earth." The dinosaurs would have appeared around 22:46, and ruled Earth for the next 50 minutes. Our species did not evolve until three seconds before midnight, left Africa a second later, and built the first cities a tenth of a second before midnight.

If the history of the planet were encompassed in the width of your outstretched arms, the distance from the fingertips of your left hand to the wrist on your right hand would be the Precambrian period. Complex life only takes up the space of your right hand, while one shave of a nail file could eradicate human history.

If you traveled back in time at a rate of **one year per second**, it would take half an hour to reach Jesus and just over three weeks to visit the dawn of humanity. It would take 23 years to reach the first complex animals, and 145 years to witness the formation of Earth.

If the entire history of the universe were compressed into **a year**

The Big Bang would have occurred on New Year's Day

The Sun and Earth would have formed in August

The first life would have appeared in September

The dinosaurs would have ruled from December 24th–29th

Modern humans would have appeared at 23:54 on December 31st

Columbus would have sailed the ocean blue one second before midnight

J F M A M J J A S O N D

The geologic clock

PLEISTOCENE
(2.58 mya) 23:59:10

PLIOCENE
(5.33 mya) 23:58:20

MIOCENE
(23.03 mya) 23:52:22

OLIGOCENE
(33.9 mya) 23:49

EOCENE
(55.8 mya) 23:42

PALEOCENE
(65.5 mya) 23:39

CRETACEOUS
(145.5 mya) 23:14

JURASSIC
(199.6 mya) 22:56

TRIASSIC
(251.0 mya) 22:40

PERMIAN
(299.0 mya) 22:24

CARBONIFEROUS
(359.2 mya) 22:05

DEVONIAN
(416.0 mya) 21:47

SILURIAN
(443.7 mya) 21:38

ORDOVICIAN
(488.3 mya) 21:23

APE-HUMAN ANCESTORS (7 mya) 23:57
BIRDS (147 mya) 23:14
MAMMALS (225 mya) 22:49
DINOSAURS (235 mya) 22:46
REPTILES (340 mya) 22:13
SEMI-AQUATIC VERTEBRATES (360 mya) 22:07
LAND ANIMALS (418 mya) 21:46
LAND PLANTS (475 mya) 21:31
EUKARYOTES (app. 1.85 bya) 15:00
LIFE ON EARTH (app. 3.8 bya) 04:10

Cenozoic

Mesozoic

Paleozoic

Hadean

Archean

Proterozoic

The Moon was created at about eight minutes past midnight, when the very young Earth was impacted by a planet-sized body named Theia, blasting free material to create our satellite.

24 hrs = 4,500 million years
1 hr = 187.5 million years
1 min = 3.125 million years
1 sec = 52 thousand years

Is evolution a bush or a tree?

▶ Darwin described the evolutionary history of life as being like a branching tree, but it is more accurate to call it a bush.

In *On the Origin of Species*, Darwin famously compared the history of life to a tree: "Of the many twigs which flourished when the tree was a mere bush, only two or three now grow into great branches, yet survive and bear the other branches." Darwin was a careful thinker who chose his words with great precision, but the Victorian world that came to embrace his theory tended to interpret his ideas in ways that suited its preconceptions. The bush element of his analogy was ignored, and the tree was seized upon. Evolution was seen as progressive, mounting from primitive roots toward a sunlit crown, where, atop the highest branch, sat man.

In fact, the tree became more of a ladder, and it was not only plants, animals, and microscopic life that were ranked into a hierarchy; the human races were also seen in hierarchical terms, with the white man on the highest rung—the pinnacle of evolution. In this sense, it is a classic example of how analogies can mislead and be misused. It is more accurate to say that evolution is like a bush. Species adapt to their ecological niches more or less successfully, and as the ecology changes, species also change. Although speciation radiates out from common ancestors, there is no value or hierarchy implied in a species' position on the bush. Being nearer the outside of the bush simply means that a species arose more recently.

A sketch from Darwin's own notebooks shows how his visualization of an evolutionary lineage was much more bushlike than treelike.

I think

Life at the extremes: The bush analogy works best if you imagine the bush growing against an old wall, and thrusting shoots into the tiniest and most hard-to-reach cracks. Similarly, species evolve to adapt to the most extreme ecological niches. Extreme organisms are proof of the power of evolution.

Hypsibius dujardini is a species of tardigrade, a microscopic organism that can be found in the most inhospitable environments on Earth.

Tardigrades are **polyextremophilic**, meaning that they can tolerate multiple adverse conditions, including temperature **extremes**, high radiation levels, and lack of oxygen.

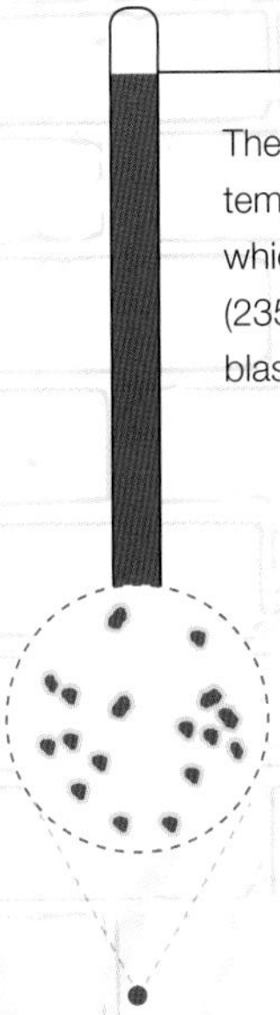

The organism that can survive the **hottest** temperature is a genus of microbes called *Pyrolobus*, which can survive and thrive at temperatures of 113°C (235°F), and may even be able to survive a ten-hour blast at 121°C (250°F).

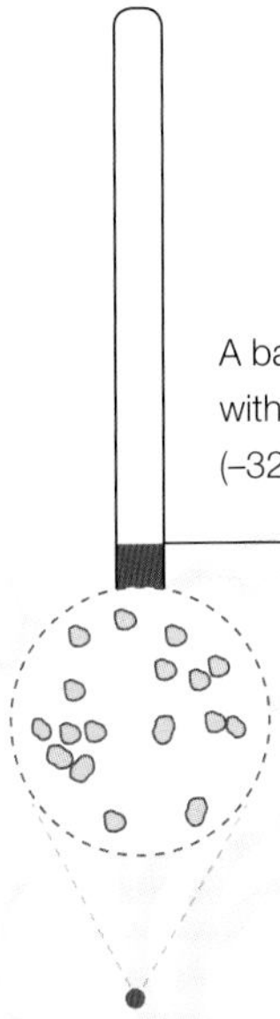

A bacterium called *Colwellia psychrerythraea* can withstand **extreme cold** as low as −196°C (−320°F)—the temperature of liquid nitrogen.

Haloarcula marismortui is a microbe that thrives in the Dead Sea, in water eight times **saltier** than the ocean.

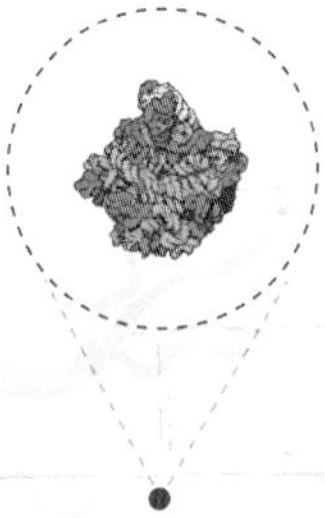

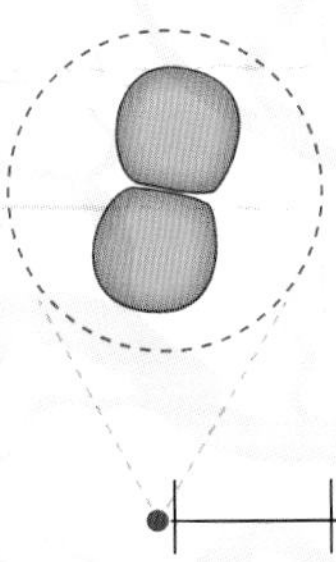

Microbes of the genus *Picrophilus* can survive **acid** as strong as pH 0–battery-acid strength.

Natronomonas pharaonis can survive conditions as **caustic** as pure ammonia (pH 12).

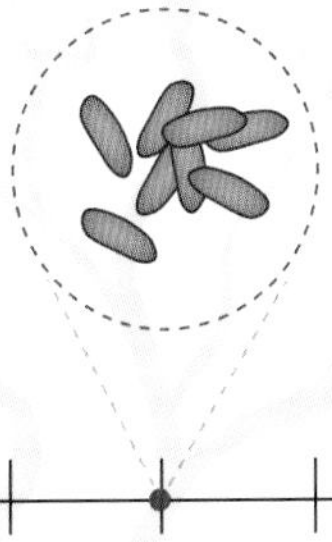

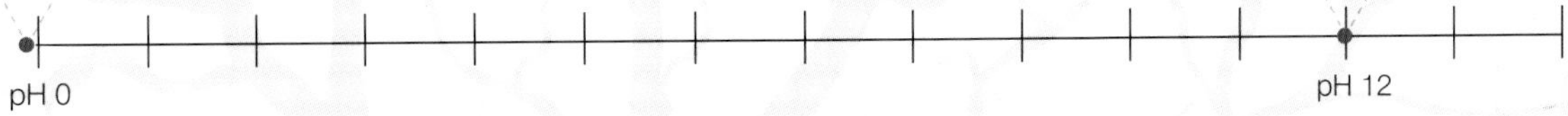

pH 0

pH 12

Microbes have been discovered in some of the deepest waters on Earth. Some may thrive at depths greater than 5 km (3 miles) below the surface of the land.

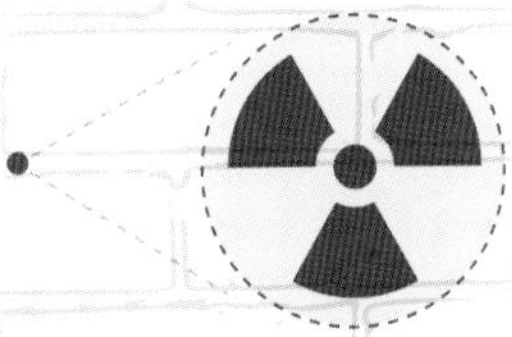

A genus of bacteria called *Geobacter* can **eat uranium**, while *Deinococcus radiodurans* can withstand radiation 2,000 times stronger than the lethal dose for a human.

Little and large

▶ If you blew up the smallest organism in the world so that it was as wide as a soccer ball, the largest organism would be two and a half times bigger than Earth.

The disparities of scale evident in the natural world are mind-boggling. The largest (ever) organism in the world, the blue whale, is over eight orders of magnitude bigger than the smallest, a microscopic bacterium isolated from a mine in California, known as ARMAN (Archaeal Richmond Mine acidophilic nano-organism). This is the sort of disparity seen between astronomical bodies and objects familiar from everyday life. Perhaps even more incredible is that the organisms at these extremes, and all the ones in between, are practically identical in terms of the molecules from which they are constructed.

The smallest and largest known organisms are shown here side by side. In reality, an ARMAN bacterium is 180 million times smaller than a blue whale.

The diver depicted here alongside the blue whale is shown to scale, making clear the disparity in size between Earth's largest ever organism and a human being.

Meters

0 1 2 3 4 5 6 7 8 9 10 11 12 13 14 15 16 17 18 19 20 21 22 23 24 25 26 27 28 29 30

180

The blue whale, *Balaenoptera musculus*, is 30 m (100 ft) long and weighs

tonnes or more

They can live for over **100 years**

They can swim at up to **32 km/h** (20 mph)

A blue whale's **tongue** weighs as much as an elephant and its heart is the size of an automobile. Some of its blood vessels are wide enough for a man to swim along. Its intestines are 120 m (390 ft) long.

They can make noises **so loud** they can be heard up to

1,600 km (1,000 miles) away.

That's nearly **1/25** of Earth's circumference.

4

tonnes is the amount of tiny shrimplike creatures called krill a blue whale can eat in a single day.

ARMAN bacteria are **200 nanometers** (0.0002 mm) long, about ⅓ the size of the common lab bacterium *E. coli* and 150 million times shorter than a blue whale.

ARMAN

Its genome is **3,000** times smaller than the human genome.

Even smaller than ARMAN are viruses, but many scientists do not consider these to be living organisms in the same sense. The smallest virus is just **20 nanometers**, 1.5 billion times shorter than a blue whale.

Virus

The fungus three times the size of Central Park

▶ The largest collective organism in the world is the size of 1,665 football fields, three times bigger than Central Park in New York.

What constitutes a single organism? If it is a collection of genetically identical cells that communicate and are coordinated for a common biological purpose, the definition of an organism must include mycorrhizal networks: mats of tiny, threadlike "roots" that send up fruiting bodies (mushrooms). In this case, the largest organisms in the world are the vast mycorrhizal networks that extend across huge swathes of forest floor.

Clones are also genetically identical, which raises interesting questions about measuring longevity. If an amoeba splits into two genetically identical cells, are they both daughter cells or is one the "parent" and the other the "offspring"? If the amoeba is simply said to have budded off new individuals while itself remaining essentially the same organism, then amoebae and other organisms that reproduce by cloning can be said to be effectively immortal.

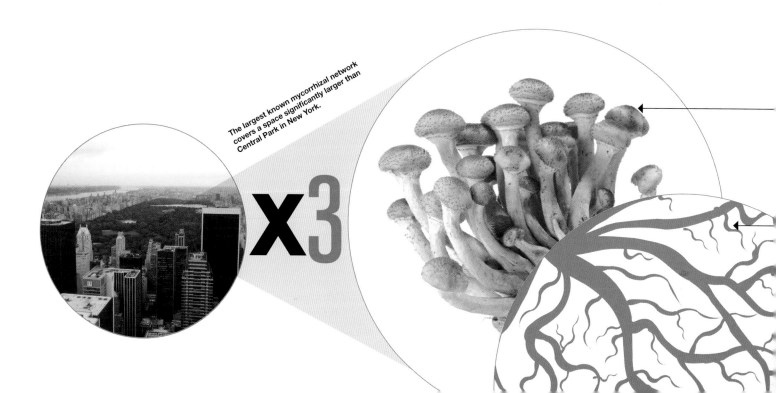

The largest known mycorrhizal network covers a space significantly larger than Central Park in New York.

x3

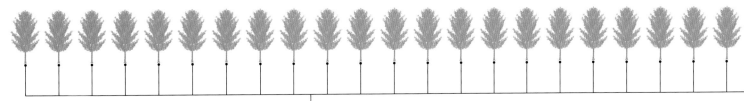

In Utah there is a grove of genetically identical quaking aspen trees that share a **single root system**. This clonal colony, named Pando, may be the heaviest organism in the world, weighing 6,000 tonnes—equivalent to one-quarter the mass of the *Titanic*, or 33 blue whales.

At **80,000** years old, Pando may also be the oldest organism in the world.

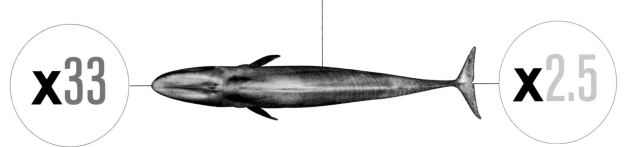

x33

x2.5

A **mushroom farm** can produce 454 tonnes (1 million pounds) of mushrooms a year—these will all be genetically identical, so the farm could be said to be growing a 454-tonne organism every year, equivalent to cultivating 2.5 blue whales or 80 elephants a year.

The largest collective organism discovered so far is a mat of honey mushrooms (*Armillaria solidipes*) made up of genetically identical cells, which infests the Blue Mountains, a forested area of eastern Oregon.

The **mycorrhizal network** of this individual occupies 965 hectares (2,384 acres) of soil, 16 times the total floor area of the world's largest office building, the Pentagon.

It could be one of the oldest organisms on Earth at

8,650 years old

Small but deadly

▶ A pop can full of botulinum toxin could kill every human on the planet.

A poison is defined as a substance that, even in small doses, causes injury or death by means of biochemical action. In the USA the legal definition of a poison is a substance that is lethal at doses of 50 milligrams per kilogram of body weight or less. Given an adult male weight of about 70 kg (154 lb), this means that a lethal dose would be anything less than 3.5 g (0.12 oz), or about ¾ of a teaspoon.

Many organisms have evolved to be poisonous or venomous. Often this is a form of defense, to discourage predation by other organisms. Sometimes poison is a form of attack, used to immobilize, kill, and digest victims. Sometimes organisms become poisonous by accident—by consuming large amounts of toxic microorganisms, for example, so that the toxin becomes concentrated to dangerous levels.

The most potent poison known to man is the botulinum toxin produced by the bacterium *Clostridium botulinum*. It produces seven different toxins, the most lethal of which stops nerve cells from firing, completely blocking nerve function.

Lethal doses are measured in milligrams per kilogram—in other words, to be lethal in around 50% of cases, how many milligrams of poison are needed per kilogram of victim. Botulinum toxin has a lethal dose of just 1 nanogram (0.000001 mg/kg), meaning that to kill someone of average weight you'd need

The lethality of botulinum toxin differs according to which source you consult—possibly because definitive experiments are difficult to conduct with such a deadly poison.

70ng

This means that just 470 g of toxin—less than the weight of a pint of milk—would be enough to kill over 6 billion people.

Arrow frogs get their name from their use by Amerindians as a source of poison for arrows and darts. They secrete toxins including batrachotoxin, one of the most potent toxins in nature. Batrachotoxin has a lethal dose of less than 200 micrograms— ½ the weight of a snowflake.

X3

According to some sources the lethal dose of botulinum toxin may be as low as 0.1 ng/kg, in which case around three tablespoons would be

enough to kill the entire human race

Possibly the most venomous animal in the world is the Australian sea wasp, or box jellyfish. Each jellyfish contains enough poison to kill 60 humans, in as little as

4 minutes

1 bite

The snake with the most toxic venom is the inland taipan, *Oxyuranus microlepidotus*, also known as the fierce or small-scaled snake. A single bite from the inland taipan can contain enough venom to **kill 100 human adults** or up to 250,000 mice. Fortunately, it is timid and docile unless provoked.

The inland taipan is over 50 times more toxic than a king cobra.

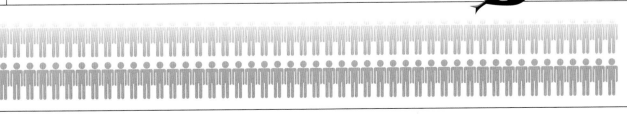

Hot, smoky, and irreverent pain

▶ According to Dr. Justin Schmidt, being stung by a yellow jacket wasp is like W. C. Fields putting out a cigar on your tongue—hot, smoky, and irreverent.

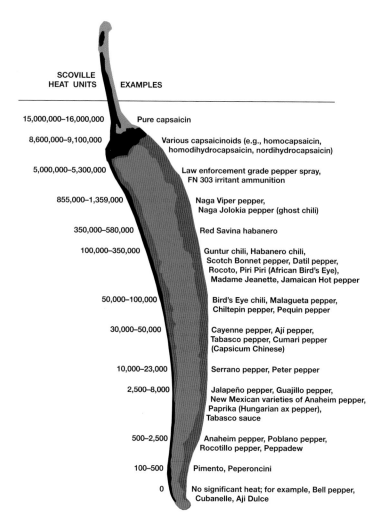

SCOVILLE HEAT UNITS	EXAMPLES
15,000,000–16,000,000	Pure capsaicin
8,600,000–9,100,000	Various capsaicinoids (e.g., homocapsaicin, homodihydrocapsaicin, nordihydrocapsaicin)
5,000,000–5,300,000	Law enforcement grade pepper spray, FN 303 irritant ammunition
855,000–1,359,000	Naga Viper pepper, Naga Jolokia pepper (ghost chili)
350,000–580,000	Red Savina habanero
100,000–350,000	Guntur chili, Habanero chili, Scotch Bonnet pepper, Datil pepper, Rocoto, Piri Piri (African Bird's Eye), Madame Jeanette, Jamaican Hot pepper
50,000–100,000	Bird's Eye chili, Malagueta pepper, Chiltepin pepper, Pequin pepper
30,000–50,000	Cayenne pepper, Ají pepper, Tabasco pepper, Cumari pepper (Capsicum Chinese)
10,000–23,000	Serrano pepper, Peter pepper
2,500–8,000	Jalapeño pepper, Guajillo pepper, New Mexican varieties of Anaheim pepper, Paprika (Hungarian ax pepper), Tabasco sauce
500–2,500	Anaheim pepper, Poblano pepper, Rocotillo pepper, Peppadew
100–500	Pimento, Peperoncini
0	No significant heat; for example, Bell pepper, Cubanelle, Aji Dulce

Justin Schmidt is an entomologist who specializes in the study of the Hymenoptera, the order of insects that includes bees, wasps, and ants, many species of which can sting for defense or attack. Personal experience led Schmidt to formulate a scale rating the relative painfulness of the stings of different species, and describing the sensations with inventive analogies.

The Schmidt scale is purely subjective. An attempt to produce a more objective scale to rate the heat of chilies was made by American chemist Walter Scoville. Scoville produced very high dilutions of the chemical responsible for giving chilies their heat—capsaicin—and fed these to a panel of five tasters. The dilution at which the tasters could detect heat gave the chili its Scoville rating. This method was still partly subjective, and more recently a machine that counts the actual concentration of capsaicin has been used to provide a completely objective scale measured in American Spice Trade Association pungency units, which are roughly convertible to Scoville units by multiplying by 15.

The Scoville Scale is the chili lover's bible, but has actually been superseded by the completely objective American Spice Trade Association pungency scale.

The **Schmidt Sting Pain Index**

1.0 Sweat bee:
Light, ephemeral, almost fruity. A tiny spark has singed a single hair on your arm.

1.2 Fire ant:
Sharp, sudden, mildly alarming. Like walking across a shag carpet and reaching for the light switch.

1.8 Bullhorn acacia ant:
A rare, piercing, elevated sort of pain. Someone has fired a staple into your cheek.

2.0 Bald-faced hornet:
Rich, hearty, slightly crunchy. Similar to getting your hand mashed in a revolving door.

2.0 Yellow jacket:
Hot and smoky, almost irreverent. Imagine W. C. Fields extinguishing a cigar on your tongue.

2.0 Honeybee:
Like a match head that flips off and burns on your skin.

3.0 Red harvester ant:
Bold and unrelenting. Somebody is using a drill to excavate your ingrown toenail.

3.0 Paper wasp:
Caustic and burning. Distinctly bitter aftertaste. Like spilling a beaker of hydrochloric acid on a paper cut.

4.0 Pepsis wasp:
Blinding, fierce, shockingly electric. A running hairdryer has been dropped into your bubble bath.

4.0+ Bullet ant:
Pure, intense, brilliant pain. Like walking over flaming charcoal with a three-inch nail in your heel.

A single **honeybee** carries this amount of venom.

~600 micrograms

It would take 47,000 bees—more than two hives' worth—to get 28 ml (1 fl oz) of liquid venom.

Bee venom is made up of 40 different toxins

Humans are **terrified** of stinging insects, despite generally being at least a million times bigger.

Africanized honeybees have the reputation of being "**killer bees**," but since 1990 only 11 people in the USA have died from "killer" bee stings. More people die from dog attacks each year.

In the USA about **4.7 million** people get bitten by dogs each year; most of these are not serious enough to need a visit to the doctor, but every day 1,000 people in America visit an emergency room for treatment for dog bites.

The spider that caught a jumbo jet

▶ A cable no wider than a pencil, spun from spider's silk, could stop a jumbo jet dead in its tracks.

A graphic illustration of the amazing strength of spider's silk is the often-quoted claim that a strand the width of a pencil (about 6 mm, or ¼ in) can stop a Boeing 747 "Jumbo Jet" in mid-flight. This claim is questionable (see below), but there is no doubt that spider's silk is one of the world's most remarkable substances, and a prime example of how, through the power of evolution, materials exist in nature that rival or outperform the best that human technology can contrive.

A spider's silk is composed of long, sticky molecules woven together into a fiber. A spider can produce different types of silk for different purposes. The strongest silk is "dragline" silk, which the spider uses as a safety line. The properties of dragline silk depend on its water content and the species producing it, but most sources agree that its tensile strength (the amount of stretching it can take before breaking) is greater than most types of steel and comparable with Kevlar, one of the strongest artificial fibers (used for bulletproof vests).

> 6.4 mm

If a spider could spin a strand of silk 6.4 mm (⅜ in) thick and 30 km (19 miles) long, it could catch a jumbo jet. The spider in question, however, would have to be 128 m (420 ft) across.

Golden orb spiders (genus *Nephila*) have the **thickest silk** (0.01 mm thick) and the biggest webs (2 m, or 6 ft, in diameter).

The **largest** webs are built by communal spiders. For instance, *Ixeuticus socialis* in Australia builds webs up to **1.2** m wide and **3.7** m long

"Capture silk" is used to stop flying insects dead in their tracks. Measured in megapascals (MPa), the tensile strength of capture silk is

1,338 MPa

According to calculations by Ed Nieuwenhuys and Leo de Cooman, a pencil-thick strand of dragline silk could stop a jumbo jet at landing speed (260 km/h or 160 mph) if it were **30 km long**.

Spinning a strand this long would require

x102 billion gardenspiders

The tensile strength of "mild" steel is

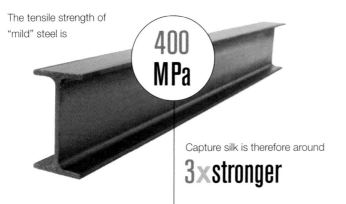

400 MPa

Capture silk is therefore around

3x stronger

The **dragline** of a European garden spider (*Araneus diadematus*) can support a mass of (0.5 g) 0.02 oz without snapping, whereas a steel strand of similar thickness will snap under the strain of just 0.01 oz (0.25 g).

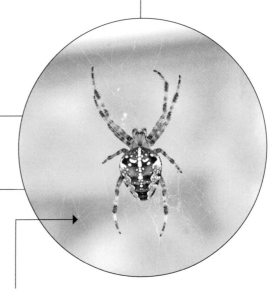

A dragline would have to be 80 km (50 miles) long before it would break under its own weight.

Flexible linkages

Hydrogen bond

Crystalline segment

The structure of spider silk helps to make it strong. Flexible "linkages" are arranged between hard crystalline segments to provide high tensile strength. At a molecular level, hydrogen bonds also contribute to its strength.

A typical strand of garden spider silk has a diameter of **0.003 mm** (0.00012 in).

Silkworm silk (used in textiles) is ten times thicker, with a diameter of **0.03 mm** (0.0012 in), but less than half as strong.

Spiders use silk as "parachute lines," to pick up gusts of wind and fly away as if attached to a kite. Spiders have been found 4,500 m (14,000 ft) above sea level and also in the middle of the ocean **1,500 km** (1,000 miles) from the nearest land.

The cell is like a city

▶ A cell is like a microscopic city, with power stations (mitochondria), factories (ribosomes), garbage trucks (excretory vacuoles), and even a city wall (cell membrane).

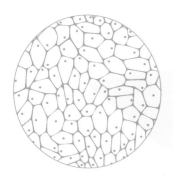

Living tissue is made up of cells—tiny compartments of biological activity, each containing a complete copy of the organism's genome, and each packed with the equipment needed to survive and carry out specific functions.

When the great 17th-century scientist Robert Hooke became the first man to describe cells, he was looking through a crude microscope at a piece of cork. The row upon row of tiny, empty compartments he saw reminded him of the cells inhabited by monks at a monastery. He and his successors soon found that all living things consisted of cells, but thanks to 400 years of increasingly sophisticated investigation, it is now known that the cell is very far from the empty space seen by Hooke. In fact, it is a fantastically sophisticated and complex assemblage of different materials and structures, busily and constantly engaged in an incredible diversity of tasks—much like a city.

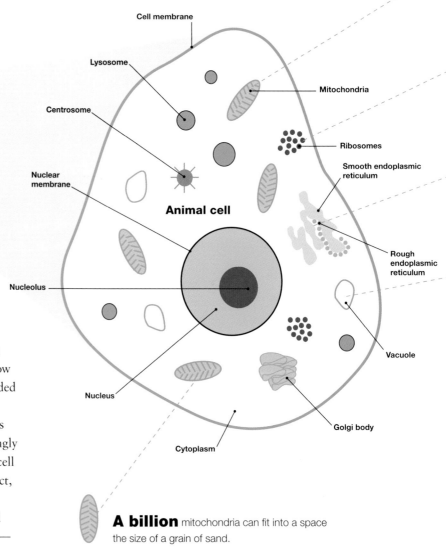

Animal cell

Cell membrane
Lysosome
Centrosome
Nuclear membrane
Nucleolus
Nucleus
Cytoplasm
Mitochondria
Ribosomes
Smooth endoplasmic reticulum
Rough endoplasmic reticulum
Vacuole
Golgi body

A billion mitochondria can fit into a space the size of a grain of sand.

Power stations
Mitochondria provide the chemical energy needed by cells to function.

Factories
Ribosomes turn amino acids into protein—the building blocks of life.

Transport system
The endoplasmic reticulum is an extensive network of sacs that has multiple functions, including the transportation of synthesized proteins around the cell.

Waste system
Vacuoles export unwanted substances from the cell.

Just as a city has power stations and factories, streets and sewers, mail services and a central government, so too a cell has mitochondria (energy-producing bodies) and ribosomes (protein-manufacturing structures), endoplasmic reticula (active transport channels), and excretory vacuoles (tiny membrane-enclosed sacs that carry waste out of the cell), signaling molecules and a nucleus (the body that contains the genetic material which coordinates the activities and outputs of the cell). Analogs can be found for almost any activity or institution in the city, but it is important to stress the limits of the analogy—cities may grow but they rarely replicate, while the cell arguably lacks anything analogous to residents.

140 trillion

The number of cells in the human body. Starting with a single cell that divides to give two cells, it would take just **47 doublings** to arrive at this number.

To **build** a very basic cell—for example, a yeast—requires more components than are found in a Boeing 777, squeezed into a space 5 microns wide (that's over 12½ million times smaller than a Boeing 777).

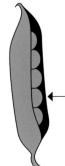

If **atoms** were magnified to the size of peas, a cell would be

~ 0.8 km wide

A typical human cell contains 20,000 different types of protein; around 2,000 of these are present in quantities of > 50,000 molecules. So there are at least

100 million protein molecules in a cell.

Supermarket sweep

▶ The natural world is organized much like a supermarket, with organisms belonging to species, which belong to families, which belong to phyla, just as products sit on shelves in aisles within departments.

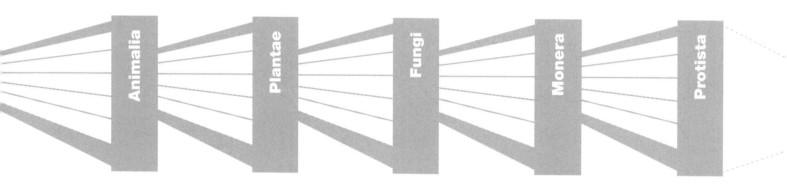

Taxonomy can be taxing. Humans belong to the genus and species *Homo sapiens*, of the family Hominidae, order Primates, class Mammalia, phylum Chordata, kingdom Animalia. This taxonomical description follows a classification arranged hierarchically, just as the products in a supermarket are arranged according to a hierarchical scheme.

The supermarket is divided into broad departments—for example, fresh produce, frozen foods, household—and these in turn are broken down into smaller departments (for example, fresh produce might consist of fruit and vegetables, dairy, meat, and fish). Each of these is further divided into aisles. Along the aisles groups of one type of product occupy the same shelf, and are arranged by brand. The brands correspond to the species, product type to genus, shelf to family, aisle to order, sub-department to class, and so on.

According to traditional taxonomy, the highest order of classification is the kingdom, and biologists used to divide all life into five kingdoms: Animalia (animals), Plantae (plants), Fungi, Monera (bacteria), and Protists (single-celled protozoans). Dramatic advances in both taxonomy and genetics, however, have led to significant revisions to this traditional scheme. For instance, it is now recognized that the Monera actually constitute an entirely different domain of life (the Prokaryota) to the other four kingdoms, which belong to the domain Eukaryota.

Microbes account for at least **80%** of global biomass.

Although impossible to say with accuracy, it is estimated that the **total** number of species that have ever lived is somewhere in the range of:

30billion**–4,000**billion

15,000 new species are recorded each year. According to the *Wall Street Journal*, there are *c.*10,000 taxonomists active in the world, at a cost of

$2,030per**species**

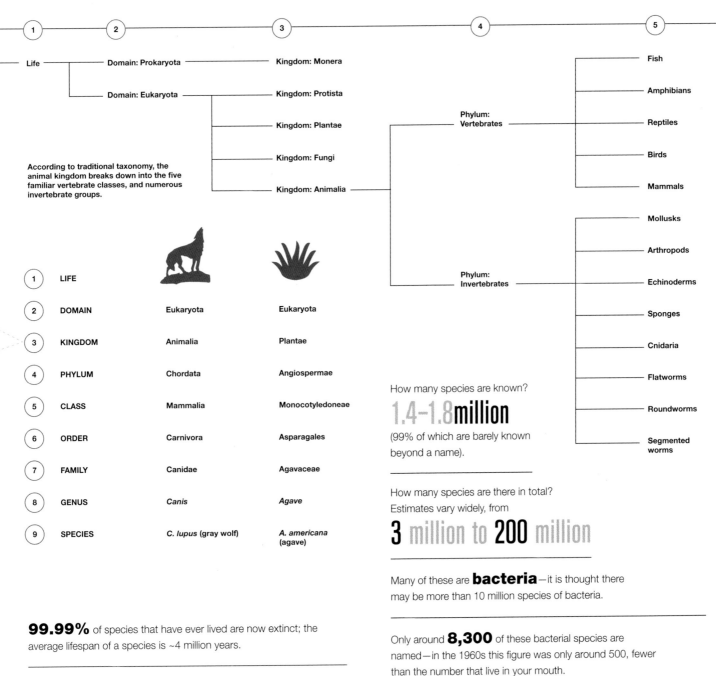

		Eukaryota	Eukaryota
①	LIFE		
②	DOMAIN	Eukaryota	Eukaryota
③	KINGDOM	Animalia	Plantae
④	PHYLUM	Chordata	Angiospermae
⑤	CLASS	Mammalia	Monocotyledoneae
⑥	ORDER	Carnivora	Asparagales
⑦	FAMILY	Canidae	Agavaceae
⑧	GENUS	*Canis*	*Agave*
⑨	SPECIES	*C. lupus* (gray wolf)	*A. americana* (agave)

1 — Life — Domain: Prokaryota — Kingdom: Monera

Domain: Eukaryota — Kingdom: Protista / Kingdom: Plantae / Kingdom: Fungi / Kingdom: Animalia

According to traditional taxonomy, the animal kingdom breaks down into the five familiar vertebrate classes, and numerous invertebrate groups.

Phylum: Vertebrates — Fish / Amphibians / Reptiles / Birds / Mammals

Phylum: Invertebrates — Mollusks / Arthropods / Echinoderms / Sponges / Cnidaria / Flatworms / Roundworms / Segmented worms

How many species are known?

1.4–1.8 million

(99% of which are barely known beyond a name).

How many species are there in total? Estimates vary widely, from

3 million to 200 million

Many of these are **bacteria**—it is thought there may be more than 10 million species of bacteria.

Only around **8,300** of these bacterial species are named—in the 1960s this figure was only around 500, fewer than the number that live in your mouth.

99.99% of species that have ever lived are now extinct; the average lifespan of a species is ~4 million years.

The background rate of extinctions throughout the history of life has been roughly one species every four years; the rate of human-caused extinction is approximately 120,000 species every four years.

3/23

Of the 23 major divisions of life, only three are visible to the naked eye.

Enzymes hold the keys to biochemical success

▶ An enzyme acting on its substrate is like a key that fits precisely into a lock and unlocks it.

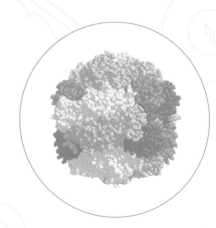

An enzyme is a biological catalyst that welds together or breaks apart other molecules without being changed itself. Life depends on enzymes—human metabolism, for instance, is entirely dependent on the activity of enzymes. Enzymes get their almost magical abilities from their unique shapes.

They are long, complex protein molecules that are folded and twisted to produce pockets that are precisely fitted to the enzymes' substrates—the molecules that they target. These pockets, or active sites, are like the tines and notches of a key; just as the notches are shaped precisely to fit into the barrel of a lock, the active site of the enzyme fits precisely around the contours of the substrate, or part of it.

Just as turning a key causes tumblers to move, unlocking a lock, the active site of the enzyme causes chemical changes to the target site of the substrate, making it break apart or join together with another substrate molecule. And just as the key can be withdrawn unchanged, and used to open other locks, the enzyme is not altered and can move on to perform the same operation on other substrates.

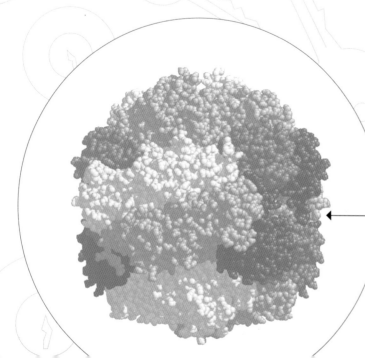

The most abundant molecule on Earth is an enzyme: RuBisCO, or ribulose-bisphosphate carboxylase oxygenase, which plays a central role in photosynthesis. There may be around 40 million tonnes of Rubisco in the biosphere—about 8 kg (18 lb) per person on the planet.

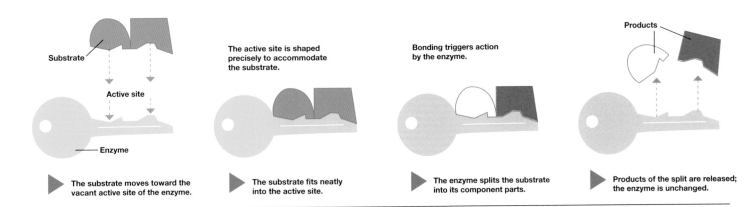

Substrate / **Active site** / **Enzyme**

The substrate moves toward the vacant active site of the enzyme.

The active site is shaped precisely to accommodate the substrate.

The substrate fits neatly into the active site.

Bonding triggers action by the enzyme.

The enzyme splits the substrate into its component parts.

Products

Products of the split are released; the enzyme is unchanged.

Enzymes are fundamental to the processes and functions of biological cells. They are able to carry out up to

1,000 tasks per second

Enzymes are biological catalysts of **amazing power**. For instance, by using enzymes soil bacteria are able to fix nitrogen into ammonia even in cold weather. To achieve the same reaction in human industry, the reactants must be subjected to 500°C (932°F) temperature and pressure of 300 atmospheres.

70x

Enzymes can be disrupted by **high salt levels**. For instance, sea water contains 70 times more salt than the human body can metabolize.

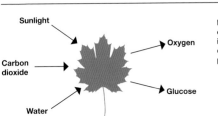

Sunlight / Carbon dioxide / Oxygen / Water / Glucose

Photosynthesis converts carbon dioxide and water into glucose, emitting oxygen as a waste product.

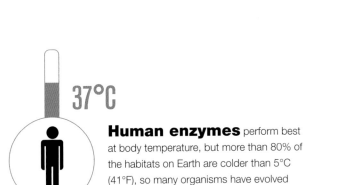

37°C

Human enzymes perform best at body temperature, but more than 80% of the habitats on Earth are colder than 5°C (41°F), so many organisms have evolved low temperature enzymes.

Per year, **photosynthesis** in marine microorganisms pumps out

150 million tonnes of oxygen

Blueprint for life

▶ The DNA of an organism is like the blueprint for a machine, or the recipe for a meal, or the code for a computer program.

DNA is the genetic material used by almost every organism on Earth. However, organisms are not made of DNA—they are made of proteins, fats, and carbohydrates, which in turn are assembled and marshaled by other proteins. In other words, proteins are the ones doing all the real "work" in an organism. So what is the link between the hereditary material that gets passed from parent to offspring (the genome), and the proteins that actually comprise these individuals (the proteome)? Somehow, DNA gives instructions for making proteins and directing their activities—the genome must translate these to the proteome.

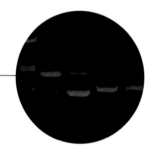

The genome can be cut up into short chunks of DNA, which can be separated using gel electrophoresis to give a distinctive pattern, known as a DNA fingerprint. Because each person's genome is unique, so is their DNA fingerprint.

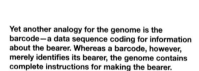

Yet another analogy for the genome is the barcode—a data sequence coding for information about the bearer. Whereas a barcode, however, merely identifies its bearer, the genome contains complete instructions for making the bearer.

DNA is made up of long sequences of molecules called nucleotide bases, which come in four primary "flavors"—adenine (A), cytosine (C), guanine (G), and thymine (T). When DNA was first discovered, it was assumed that these four types of base were insufficient for spelling out the instructions for the immense complexity of a whole organism. It would be like trying to write the works of Shakespeare with just four letters: A, C, G, and T. In fact it is entirely possible to achieve this, if these four letters are used to create a code. Once biologists realized this, it did not take them long to crack the code. Proteins are composed of 20 different amino acids, so what is needed is a code capable of at least 20 different messages, or "codons." With four options at each letter, sequences of two nucleotides would give just (4 x 4 =) 16 different two-letter words or codons. But with sequences of three nucleotides, it is possible to spell out (4 x 4 x 4 =) 64 different codons— more than enough to specify all 20 amino acids.

The DNA of a humble single-celled **amoeba** contains as many as

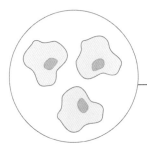

400 million

bits of genetic information—enough to fill 80 500-page books.

Every cell in your body (except red blood cells) contains a DNA sequence 3.2 billion letters long—that's

2 | meters | DNA | 👤

A 1-millimeter stretch of DNA contains a base pair sequence of more than **3 million** letters.

It would take a person typing

60	words per minute
8	hours a day
50	years

to type the human genome

8.4

Chemicals and other agents frequently attack and damage the DNA in a human cell. Such events can occur as often as every

seconds or 10,000 times a day

DNA is replicated during cell division. On average, a human cell can divide once a day. Given an adequate supply of nutrients, in a single day a bacterium can divide as many times as

280 trillion

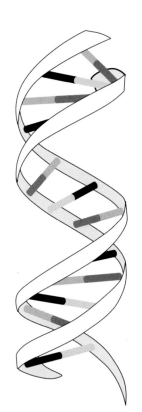

In a strand of DNA, the bases (the letters of the genome) are just 0.33 nanometers apart—about three times the width of a hydrogen atom.

A bacterium such as *Clostridium perfringens* can reproduce once every **nine minutes.** Theoretically at this rate a single bacterium could produce more offspring in two days than there are protons in the universe.

The genome as a journey

▶ If the entire length of your genome were equivalent to the distance between New York and Chicago, there would be 3 million letters per mile and a third of the distance would be covered in repeats of the same message.

British geneticist Steve Jones came up with an analogy to get across the nature of the genome—in particular the way that much, if not most, of it is made up of apparent "junk" DNA, with the actual genes few and far between. He suggested that if the entire human genome were imagined as a journey of about 1,600 km (1,000 miles), it would stretch from Land's End to John O'Groats (the southwest and northeast corners of Britain) or from New York to Chicago, with 50 bases (or letters) per inch, and 3 million per mile.

Following such a route across Britain would take you through 23 counties, corresponding to the 23 chromosomes of the human genome. Along most of the journey there would be little of note to see, with long stretches where nothing seemed to be happening, and where the roadside was monotonous. This corresponds to long sequences of repeats of the same sequence of letters, which fill about one-third of the genome. Occasionally you would pass through a town, full of activity and purpose, corresponding to a gene. Cities correspond to families of genes. Alongside these working genes are abandoned and decrepit ruins—the relics of our evolutionary past, "fossil" genes that no longer perform useful functions but get passed on anyway.

Chicago

New York

The journey from New York to Chicago takes in many interesting landmarks and busy towns and cities, but also long stretches of boring landscapes and dull strip malls. Similarly, the human genome has areas of interest and activity (genes) studded around among long stretches of "junk" DNA.

1,000 miles
(1,600 km)

The **Human Genome Project** used automated machines to sequence the human genome (i.e. read the order of base pairs). The human genome was sequenced to an accuracy of 1 in 10,000—in other words, the HGP only got one letter wrong for every 10,000 letters of the sequence.

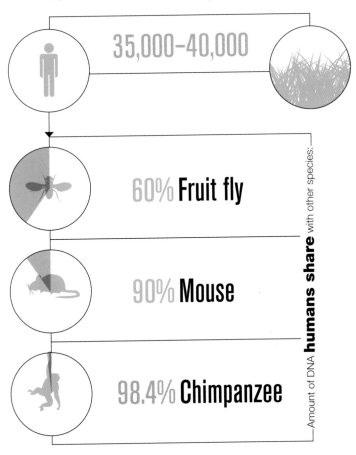

Humans have the same number of genes as grass, a total of:

35,000–40,000

Amount of DNA **humans share** with other species:

60% Fruit fly

90% Mouse

98.4% Chimpanzee

Your DNA is **99.99% identical** to every other human. Humans differ from one another by one nucleotide base in a thousand. There is greater genetic diversity in a troop of 55 chimps than in the entire human population.

How many ancestors did you inherit your DNA from? If none of your ancestors are related, going back eight generations (c.mid-19th century) you have 250 ancestors. Going back to c.1600 you have 16,384. Going back 20 generations, you have over a million ancestors, and by 30 generations over a billion.

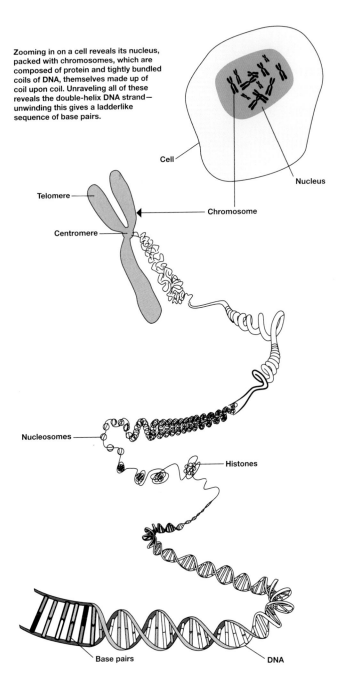

Zooming in on a cell reveals its nucleus, packed with chromosomes, which are composed of protein and tightly bundled coils of DNA, themselves made up of coil upon coil. Unraveling all of these reveals the double-helix DNA strand—unwinding this gives a ladderlike sequence of base pairs.

Cell

Nucleus

Telomere

Chromosome

Centromere

Nucleosomes

Histones

Base pairs

DNA

HIV has fewer than 10 genes; even the simplest bacteria have thousands.

Walking up the down escalator

▶ Homeostasis, the process by which organisms maintain their internal environments within acceptable parameters, is like walking up the down escalator.

Like most organisms, humans need to maintain a constant internal environment. Metabolic processes only function properly within a restricted band of temperature, pressure, salinity, water content, oxygen concentration, pH, and so on. For instance, if body temperature drops below 35°C (95°F), human physiology begins to fail and death will eventually follow.

Yet the environment around us varies constantly and is usually above or below our internal tolerance parameters; the same goes for most other organisms. Organisms square this circle through the process of homeostasis ("steady state"): when the external environment changes, they respond to maintain the internal environment within tolerance. This is similar to someone walking up the down escalator.

The human body's homeostatic response to temperature is both voluntary (for example, if you feel cold, you might put on more clothes) and involuntary (sweating and shivering are involuntary processes). Involuntary response to temperature change is controlled by the hypothalamus, a structure in the center of the brain that acts as a sort of thermostat.

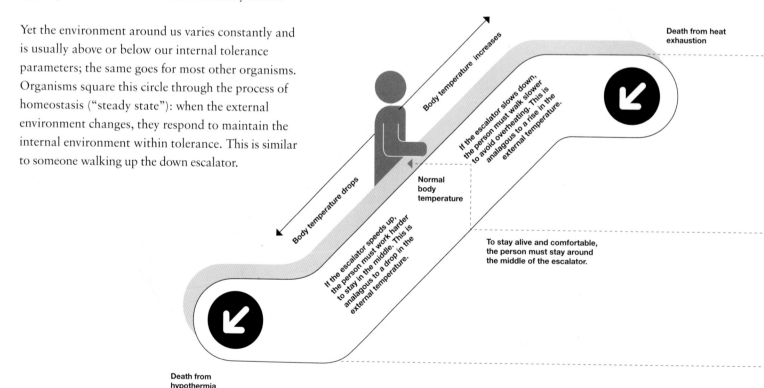

Body temperature increases

Body temperature drops

Normal body temperature

Death from heat exhaustion

If the escalator slows down, the person must walk slower to avoid overheating. This is analagous to a rise in the external temperature.

If the escalator speeds up, the person must work harder to stay in the middle. This is analagous to a drop in the external temperature.

To stay alive and comfortable, the person must stay around the middle of the escalator.

Death from hypothermia

The descent of the escalator is like heat loss from the body; reaching the bottom represents death by hypothermia. But walking up the escalator generates heat; reaching the top represents death by heat exhaustion. To stay alive, the person must remain in the middle of the escalator. If the escalator speeds up (equivalent to the environment cooling down), the person must work harder to stay in the middle (i.e., she must increase her metabolism to generate more heat—walking faster is therefore equivalent to shivering). If it slows down (external heating), she must walk slower (equivalent to sweating and panting) to avoid overheating. So homeostasis is a dynamic, not a static process.

40°C If the external temperature rises, the body cools down through sweating. **104°F**

37°C The optimum body temperature for humans is around 37°C (98.6°F). **98.6°F**

35°C If the external temperature drops, the body generates more heat by shivering, increasing the metabolic rate. **95°F**

Relative humidity can be measured as **wet-bulb** temperature— wet-bulb temperatures above 95°F are intolerable as they exceed human ability to lose heat through sweating.

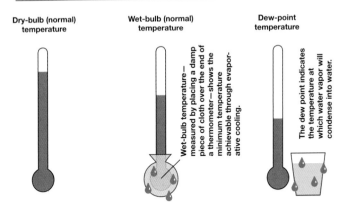

Dry-bulb (normal) temperature

Wet-bulb (normal) temperature

Dew-point temperature

Wet-bulb temperature— measured by placing a damp piece of cloth over the end of a thermometer—shows the minimum temperature achievable through evaporative cooling.

The dew point indicates the temperature at which water vapor will condense into water.

At wet-bulb temperatures of over 35°C even someone standing naked in the shade in front of a fan would **die of heat stress**.

At present, the only place on Earth where the wet-bulb temperature is **consistently beyond human tolerance** is in the Naica Cave in Mexico; for a person standing in the cave, the surface of the lungs is the coolest part of the body.

Apart from the unique, enclosed environment of the Naica Cave, virtually nowhere on Earth has a wet-bulb temperature of more than 30°C (86°F). But **global warming** could change that. If the direst predictions prove correct, with temperature rises in the tropics of 11°C (52°F), vast areas of the planet would have wet-bulb temperatures of over 35°C for at least part of the year, and would be uninhabitable.

Humans cannot long tolerate body temperatures below 95°F, yet more than 80% of the habitats on Earth can reach temperatures of **less than 41°F (5°C).**

The copepod is faster than the cheetah, and mightier than the whale

▶ If the tiny, shrimplike copepod were a human, it would accelerate to 11,500 km/h (7,160 mph) in less than a second, thanks to muscles stronger than any other species on Earth.

The cheetah is famous for being the fastest land animal, but relative to its size it is a comparative slowboat. When speed is measured in bodylengths per second (blps), a very different picture emerges, with small arthropods stealing the crown previously claimed by large vertebrates. The fastest of them all is the copepod, a tiny shrimplike creature that floats in the ocean and occasionally has to take high-speed evasive action to avoid predators. Not only are these the fastest, they are also the strongest (relative to size) and probably the most abundant multicellular animals on the planet.

Although you might think it much faster than the **domestic cat**, which has a top speed of 48 km/h (30 mph), in blps the **cheetah** (25 blps) comes in behind its domestic cousin (29 blps).

48 km/h
29 blps

0.3	2.2	6.1	25
blps	blps	blps	blps

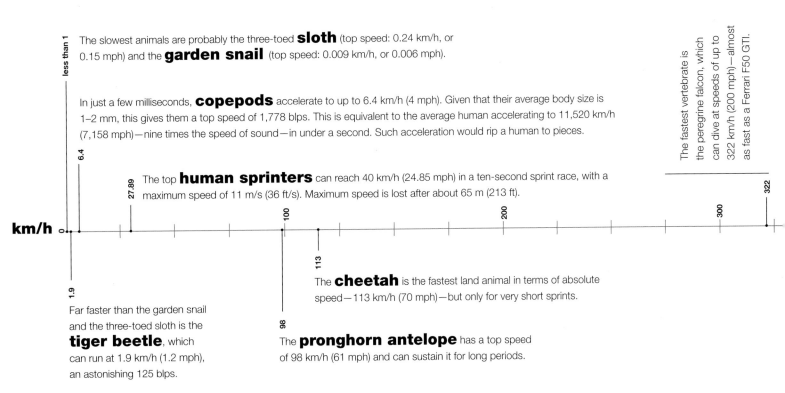

The slowest animals are probably the three-toed **sloth** (top speed: 0.24 km/h, or 0.15 mph) and the **garden snail** (top speed: 0.009 km/h, or 0.006 mph).

In just a few milliseconds, **copepods** accelerate to up to 6.4 km/h (4 mph). Given that their average body size is 1–2 mm, this gives them a top speed of 1,778 blps. This is equivalent to the average human accelerating to 11,520 km/h (7,158 mph)—nine times the speed of sound—in under a second. Such acceleration would rip a human to pieces.

The fastest vertebrate is the peregrine falcon, which can dive at speeds of up to 322 km/h (200 mph)—almost as fast as a Ferrari F50 GTI.

The top **human sprinters** can reach 40 km/h (24.85 mph) in a ten-second sprint race, with a maximum speed of 11 m/s (36 ft/s). Maximum speed is lost after about 65 m (213 ft).

km/h

less than 1 · 6.4 · 27.89 · 1.9 · 100 · 98 · 113 · 200 · 300 · 322

The **cheetah** is the fastest land animal in terms of absolute speed—113 km/h (70 mph)—but only for very short sprints.

Far faster than the garden snail and the three-toed sloth is the **tiger beetle**, which can run at 1.9 km/h (1.2 mph), an astonishing 125 blps.

The **pronghorn antelope** has a top speed of 98 km/h (61 mph) and can sustain it for long periods.

113km/h
25blps

In terms of blps, however, the peregrine falcon (200 blps) is beaten by the swallow (350 blps) and the Anna's hummingbird, which dives at speeds of up to 385 blps, making it the fastest vertebrate relative to size.

Anna's hummingbird is faster than a fighter jet with its afterburners on (150 blps), and the Space Shuttle during reentry (207 blps). As it pulls out of its dive, the hummingbird experiences **9 g acceleration**. The most a human pilot can take without blacking out is 7 g.

29
blps

207
blps

385
blps

1,778
blps

The watchmaker fallacy

▶ Walking across a blasted heath, you find a pocket watch, an intricate mechanism of gears and cogs. How did it come to be there? Did it spontaneously assemble itself, or was there a watchmaker?

?

An unlikely place to find something as complex as clockwork? Even such a barren habitat can produce extraordinary complexity through natural selection alone.

Analogies in science usually help to explain—and in some cases to inspire—scientific principles. They are recognized and recommended as powerful tools for education. But any powerful tool can be misused and abused. A classic example is the "watch on the heath" analogy, first propounded by William Paley in his 1802 book *Natural Theology*. Paley argued that a watch found on the heath could not have come to be as the result of accident and blind chance; it must have been constructed by a watchmaker. This is a version of the teleological argument for the existence of God, which says that if there appears to be some form of design evident in life/the universe, there must be a designer. This argument has been adopted by the modern Intelligent Design (ID) movement, as a way to promote Creationism.

The **watchmaker analogy** only appears reasonable because of logical fallacies and errors in reasoning. A related analogy involves being dealt a hand of cards in bridge. The odds of being dealt any specific combination of 13 cards are less than one in 600 billion. But it does not make sense to look at the hand of cards you were dealt, calculate the odds against it, and conclude that you cannot have been dealt it.

According to the ID movement, complex biological phenomena such as the human eye are analogous to the watch on the heath. The eye is said to be an "irreducibly complex" organ, in which all the components seem to work in harmony: it appears to have been perfectly designed for its purpose, which suggests that there must be some sort of intelligent designer. In fact this is a simple fallacy, confusing appearance of design with evidence of design. Evolutionary biologists have comprehensively rebutted the argument by suggesting how intermediate forms of eye could have evolved and finding fossil evidence to back this up.

Evolutionary biologists do not claim that the human eye came into being fully formed as a result of sheer chance alone, as some creationists would have you think. Like all aspects of the human body, it is the result of a long process of evolution, with many intermediate forms.

The complete **fossil legacy** of the current population of the USA will be around 50 bones—roughly **¼** of a skeleton.

1 Proponents of the ID movement point to gaps in the **fossil record** as evidence that Darwin was wrong. In fact it's amazing that there is any fossil record: just one bone out of every billion is fossilized.

bone in 1,000,000,000

1 The proportion of species that have made it into the fossil record could be as low as 250,000 out of the 30 billion that have ever lived. That's just:

species in 120,000

A similar argument lies behind the birthday paradox. Imagine you are stuck in an elevator with a handful of people, and it turns out that two of them share a birthday. This might seem like a significant quirk of fate to all concerned, but in fact the odds against such a coincidence are far lower than you might expect. To achieve a 99% probability that two people will share the same birthday, only 57 people are needed. Even with just 23 people, there is a 50% chance that two of them will share the same birthday.

Section Four

▶ Much like the microscopic scale of chemistry, the grand
scale found in astronomy also demands the use of analogy.
This section will allow the reader to understand just how
far it is to distant cosmic objects, how old the universe
really is, and how small Earth is in comparison.

Astronomy

Waltzing in the dark

▶ A black hole and a star orbiting one another in a binary system are like dancers waltzing in the dark, with one dressed in a luminous suit and the other one entirely in black.

Black holes are invisible against the blackness of space, so how do astronomers spot them? Although they cannot be observed directly, their massive gravitational power affects other astronomical objects. A star that comes too close to a black hole will be pulled into orbit around it. Astronomers can observe stars that appear to be orbiting nothing, and deduce the presence of a black hole accordingly. Similarly, if you were watching dancers waltzing in the dark but could only see one of them (because the other was entirely draped in dark cloth), you would be able to deduce the existence of the other.

In the left-hand picture both partners are visible, yet when the background is the same colour as the right-hand dancer, he disappears. Similarly, a black hole cannot be perceived directly against the blackness of space.

Black holes are created when a **star** burns itself out and the remaining matter—if there is enough of it—collapses under its own gravity into a singularity—a point with no volume and infinite density.

x30

In order to become a black hole, a star must start off at least **30 times** more massive than our Sun. Fewer than 1 out of 1,000 stars in our Milky Way are massive enough to become a black hole.

One of the closest known black holes to Earth is **Cygnus X-1**. For Earth to be sucked into this black hole it would have to be within 21 km (13 miles) of it. Thankfully, it is

6,000 **light years away**

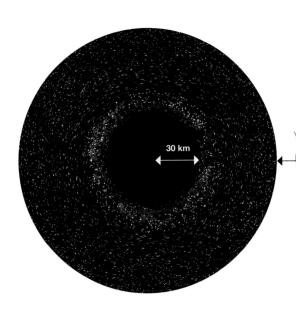

Most black holes are comparatively tiny—a typical **10-solar mass** black hole has a radius of just 30 km (18 miles)—not much bigger than London.

30 km

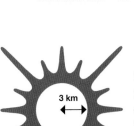

3 km

If the **Sun**—current radius 700,000 km (435,000 miles)—were replaced with a black hole of the same mass, it would have a radius of just 3 km (1.9 miles). Earth's orbit would not be affected, although it would get very dark.

Black holes at the centers of galaxies grow into **monsters**. The supermassive black hole at the center of the Milky Way is 4 million times more massive than the Sun and 24 million km (~15 million miles) across. Fortunately, it is more than

27,000 light years away

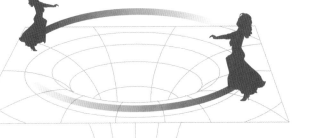

Known as a Schwarzschild embedding diagram, this is a two-dimensional graphical representation of what happens to four-dimensional space–time in the presence of a black hole.

Pulled into an orbit around a hole in space–time (i.e., a black hole), the star represented by the dancer traces a path that can be used to detect the presence of that hole.

At the center of the Perseus galaxy cluster is a **supermassive** black hole. Compared to the size of the cluster, the hole is minuscule, yet the blast waves from energy falling into it affect matter 300,000 light years away. This is like heat given off from an area the size of your fingernail affecting the entire Earth.

None more heavy

▶ A chunk of neutron star the size of a sugar cube weighs more than the human race.

Astronomical scales are difficult for the human mind to grasp, especially when dealing with such exotic objects as neutron stars, quasars, pulsars, magnetars, and white dwarfs. When a star reaches the end of its life and burns out, the explosive power of nuclear fusion no longer counteracts the pull of gravity and whatever material is left collapses under its own gravity. Stars like our Sun become very dense and shrink to roughly the size of Earth, and remain hot enough to glow dimly for the next billion years—this is a white dwarf. More massive stars have stronger gravity and crush all of their mass into the densest form of matter; even electrons are forced to merge with protons, becoming neutrons, and so these objects are called neutron stars. Rapidly rotating neutron stars that emit pulses of radiation are known as pulsars, while neutron stars with massive magnetic fields are known as magnetars. A quasar is an object not much bigger than a star that emits as much radiation as a galaxy—it is thought that quasars may be very young galaxies with supermassive black holes at their centers.

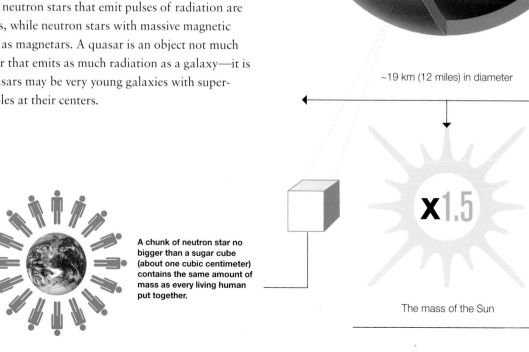

Neutron star

The solid crust is approximately 1.6 km (1 mile) thick

The super-dense interior is made up mostly of neutrons

~19 km (12 miles) in diameter

A chunk of neutron star no bigger than a sugar cube (about one cubic centimeter) contains the same amount of mass as every living human put together.

x1.5

The mass of the Sun

The material that makes up a neutron star is the same as what you would get if you were to **squeeze all the space out of an atom** (see pp. 84–5). Essentially, a neutron star is a giant atomic nucleus.

Quasars are comparatively small given their intense luminosity. Despite being smaller than our solar system, they pump out as much energy as

100 normal galaxies.

This is equivalent to **10 million million** suns.

A sugar cube-sized volume of **white dwarf** matter weighs as much as an average family car.

Gravity at the surface of a white dwarf is 100,000 times more powerful than at the surface of Earth.

If our Sun were **crushed** to the density of a neutron star, it would occupy the same volume as Mt Everest.

A white dwarf has a crust, beneath which is crystallized carbon, similar to a diamond. BPM 37093A, discovered in 2004, is a white dwarf near the constellation Centaurus made of **crystallized carbon** weighing 2.268 million trillion trillion kg (5 million trillion trillion lb)—roughly the same as the Sun. In diamond terms, that's 10 billion trillion trillion carats.

If Earth's gravity was this powerful, the top of the atmosphere would be below the level of a skyscraper, and the Empire State Building would be sticking out into space.

Gravitational time dilation means that on the surface of a neutron star time passes

x1.5 slower

than on Earth.

Pulsar B1508+55 was created in the constellation Cygnus but is currently hurtling out of the Milky Way at nearly 1,100 km/s (over 670 miles/s)—about 150 times faster than an orbiting space shuttle. At this speed, it could travel from London to New York in

5 seconds

Through a glass, curvy

▶ Looking at a distant galaxy with another galaxy in the way is like looking at a candle flame through the stem of a wineglass.

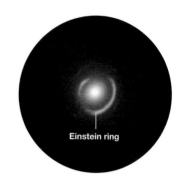

Einstein ring

Because light is bent by gravity, massive objects act like lenses in respect to light coming from behind them. By measuring the degree to which light from a distant galaxy is bent by an intervening galaxy, astronomers can actually weigh the intervening galaxy. Depending on the alignment of the two galaxies, an astronomer can obtain images of the distant one "lensed" to different degrees. If the alignment is dead on, the distant galaxy will appear as an Einstein ring around the intervening galaxy.

To see an Einstein ring for yourself, hold up a wineglass and view the flame of a candle through the flared base. Moving the base around will produce double or quadruple images. Viewing the flame directly down the stem of the glass will produce a ring image. The base of the wineglass is equivalent to a galactic-sized mass, lensing light in a similar fashion.

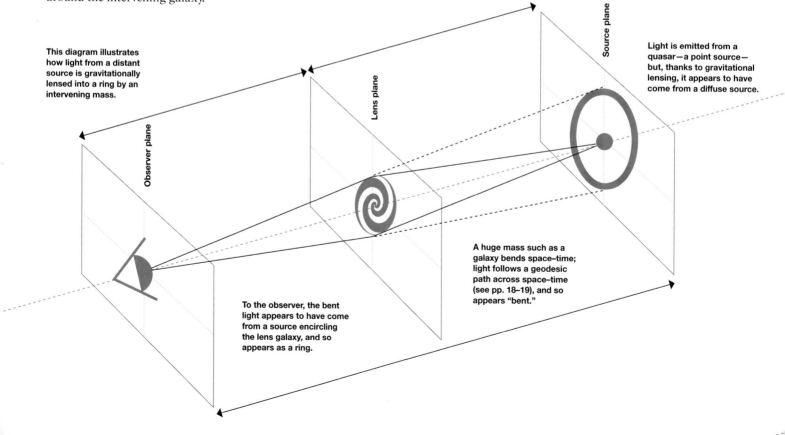

This diagram illustrates how light from a distant source is gravitationally lensed into a ring by an intervening mass.

Observer plane

Lens plane

Source plane

Light is emitted from a quasar—a point source— but, thanks to gravitational lensing, it appears to have come from a diffuse source.

To the observer, the bent light appears to have come from a source encircling the lens galaxy, and so appears as a ring.

A huge mass such as a galaxy bends space–time; light follows a geodesic path across space–time (see pp. 18–19), and so appears "bent."

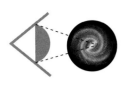

There may be up to

2 trillion

galaxies in the observable universe.

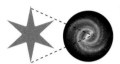

The Milky Way contains

200–400 billion stars.

The largest known galaxy, IC 1101, contains **over 100 trillion stars**

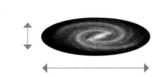

The circumference of the Milky Way is 250,000–300,000 light years.

The disk of the Milky Way is about 120,000 light years across, with a central bulge 12,000 light years across.

The longest known galaxies are up to 2 million light years across along the long axis.

The Milky Way is so full of **dust** that, in the visible spectrum, astronomers can see only 6,000 light years into the disk (that's just 6% of the diameter).

Spiral   Elliptical   Irregular

There are three main types of galaxy. Our galaxy, the Milky Way, is a spiral galaxy.

100 | In a typical galaxy, there are 10^{68} atoms. That's **million trillion trillion trillion trillion trillion**

7 | Old stars die and new ones are born all the time. NASA estimates **per year** a net gain for the Milky Way of

stars ★ ★ ★ ★ ★ ★ ★

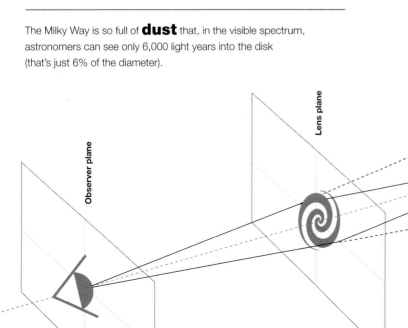

Observer plane

Lens plane

Source plane

Depending on the alignment of the source and lens, the light from the source may be lensed into a double or even quadruple image.

Inside the galactic whirlpool

▶ A spiral galaxy is like a giant whirlpool or hurricane in space. Similar formation processes are thought to be behind both earthly and cosmological phenomena.

The satellite photograph of a hurricane, left, clearly shows its spiral shape, with rings of cloud known as rain bands spiraling around a central ring known as the eye wall, itself surrounding a space known as the eye. Shown at right, a spiral galaxy has similar features—spiral arms radiating out from a central mass of dust, gas, and stars, all surrounding a void (almost certainly a supermassive black hole) at the very heart of the galaxy.

Hurricanes, whirlpools, and spiral galaxies look almost exactly alike, possibly because some of the mechanics behind the formation of each phenomenon are similar. Hurricanes and whirlpools begin with turbulence and eddies in air and water respectively. Photographs taken of the tops of clouds show many tiny eddies, and some cosmologists think that when galaxies formed in the young universe, a similar type of "turbulent priming" took place—in the vast clouds of primeval gas and dark matter, tiny local eddies grew into larger ones. Many questions remain about vortices on Earth, and vortices in space are more mysterious still, so the analogy is far from perfect. Some of the impetus for the spiral of a hurricane, for instance, comes from the Coriolis effect (see pp. 156–7), and there is no such force at work in space.

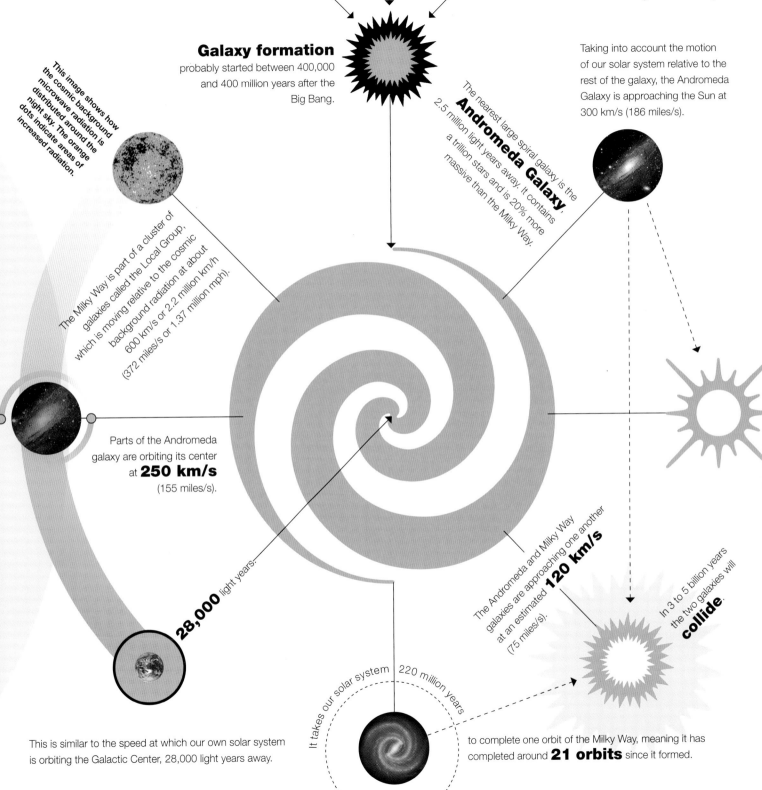

Galaxy formation
probably started between 400,000 and 400 million years after the Big Bang.

This image shows how the cosmic background microwave radiation is distributed around the night sky. The orange dots indicate areas of increased radiation.

The nearest large spiral galaxy is the **Andromeda Galaxy**, 2.5 million light years away. It contains a trillion stars and is 20% more massive than the Milky Way.

Taking into account the motion of our solar system relative to the rest of the galaxy, the Andromeda Galaxy is approaching the Sun at 300 km/s (186 miles/s).

The Milky Way is part of a cluster of galaxies called the Local Group, which is moving relative to the cosmic background radiation at about 600 km/s or 2.2 million km/h (372 miles/s or 1.37 million mph).

Parts of the Andromeda galaxy are orbiting its center at **250 km/s** (155 miles/s).

28,000 light years

This is similar to the speed at which our own solar system is orbiting the Galactic Center, 28,000 light years away.

It takes our solar system 220 million years

The Andromeda and Milky Way galaxies are approaching one another at an estimated **120 km/s** (75 miles/s).

In 3 to 5 billion years the two galaxies will **collide**.

to complete one orbit of the Milky Way, meaning it has completed around **21 orbits** since it formed.

Earth as a ball bearing

▶ If the solar system were drawn to a scale in which Earth was the size of a small ball bearing (about 2 mm wide), then the Sun would be 20 cm (8 in) wide and 23.5 meters (77 ft) away.

The universe is an unbelievably large place. Even within our own solar system, it can be difficult for us to fully comprehend the distances between the objects. When the hot Sun looms large overhead, it surely feels much closer than 150 million km (93 million miles). The Moon is our constant companion, and yet we forget that it is nearly 400,000 km (250,000 miles) away. It is only by analogy that we can begin to comprehend the true meaning of these distances. By the same scale outlined above, one would find Jupiter, less than 2.5 cm (1 in) in diameter, over 90 m (300 ft) distant. Neptune, the outermost planet in the solar system, would be 675 m (2,250 ft—more than half a mile) away from Earth, and it would be about the size of a coffee bean.

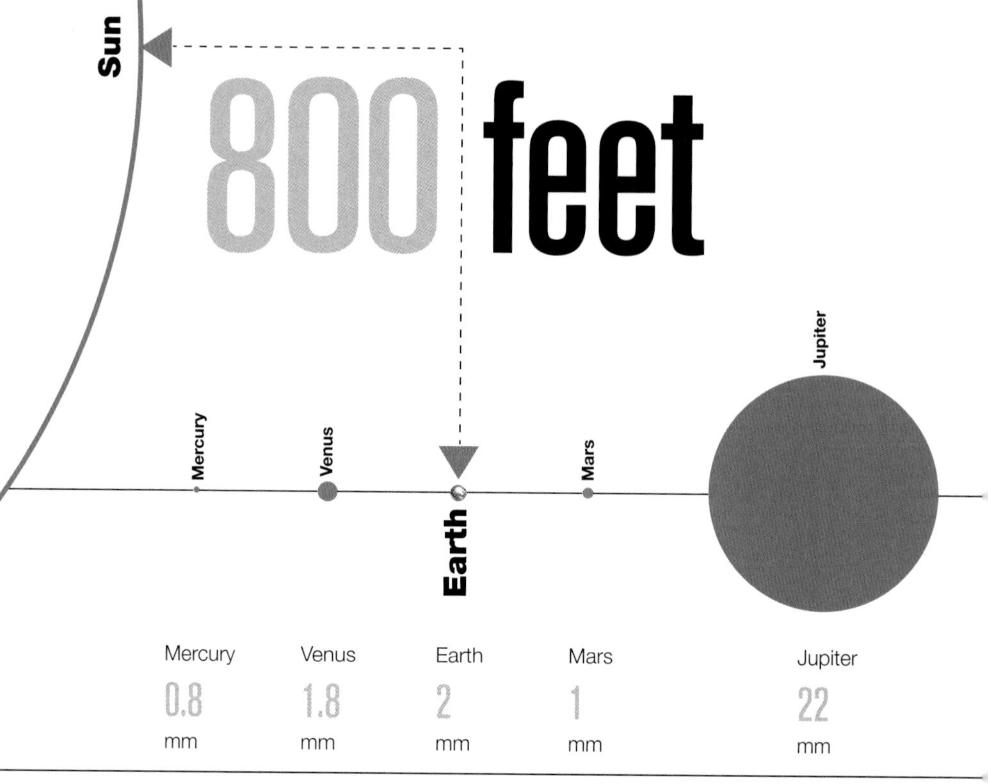

Sun

800 feet

Mercury

Venus

Earth

Mars

Jupiter

The Sun		Mercury	Venus	Earth	Mars		Jupiter
200 mm		0.8 mm	1.8 mm	2 mm	1 mm		22 mm

390,000 **kilometers**

(or 240,000 miles) is the average distance between Earth and the Moon. If Earth was the size of a basketball, the Moon would be the size of a baseball, and they would still be separated by approximately

7.6 **meters**

(or 25 feet).

8 **light**minutes

Looking beyond the solar system, while our Sun is only about

away from Earth, the distance to the next star, **Proxima Centauri**, is

4.2 **light**years

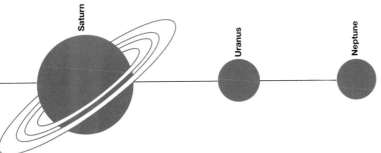

Saturn	Uranus	Neptune
18 mm	8 mm	7.8 mm

Our galaxy, the Milky Way, is one of perhaps 140 billion galaxies in the visible universe. Astrophysicist Bruce Gregory has computed that if galaxies were **peas** there would be enough to fill a large **sports arena**. This equates to as many as:

2 **trillion**

The fastest speed achieved by a human-made object (the *Helios 2* probe) is approximately **250,000 km/h** (155,000 mph). Even at that speed, the time it would take to reach the nearest star (Proxima Centauri) is more than

18 **thousand** years

13 billion

The farthest objects so far discovered in space are light bursts from distant explosions, thought to have occurred about 13 billion **light years** from Earth. This means that light from this event has been traveling through space for nearly the entire life of the universe (which is thought to be something like 13.8 billion years old). At scales such as these, even analogies become almost useless in trying to comprehend the sheer magnitude.

The **age** of the **universe** is approximately

13.8 **billion**years

A million elephants a second

▶ The Sun burns through matter at a rate equivalent to a million elephants every second.

The Sun is a giant ball of gas (mainly hydrogen and a little helium). Atoms of hydrogen are crushed together by the vast gravity of the Sun—so close that their nuclei are forced to merge, fusing to become helium nuclei. This process converts 0.8% of the mass of the hydrogen nuclei into energy—making the Sun shine brightly enough to heat Earth, 150 million km (93 million miles) away.

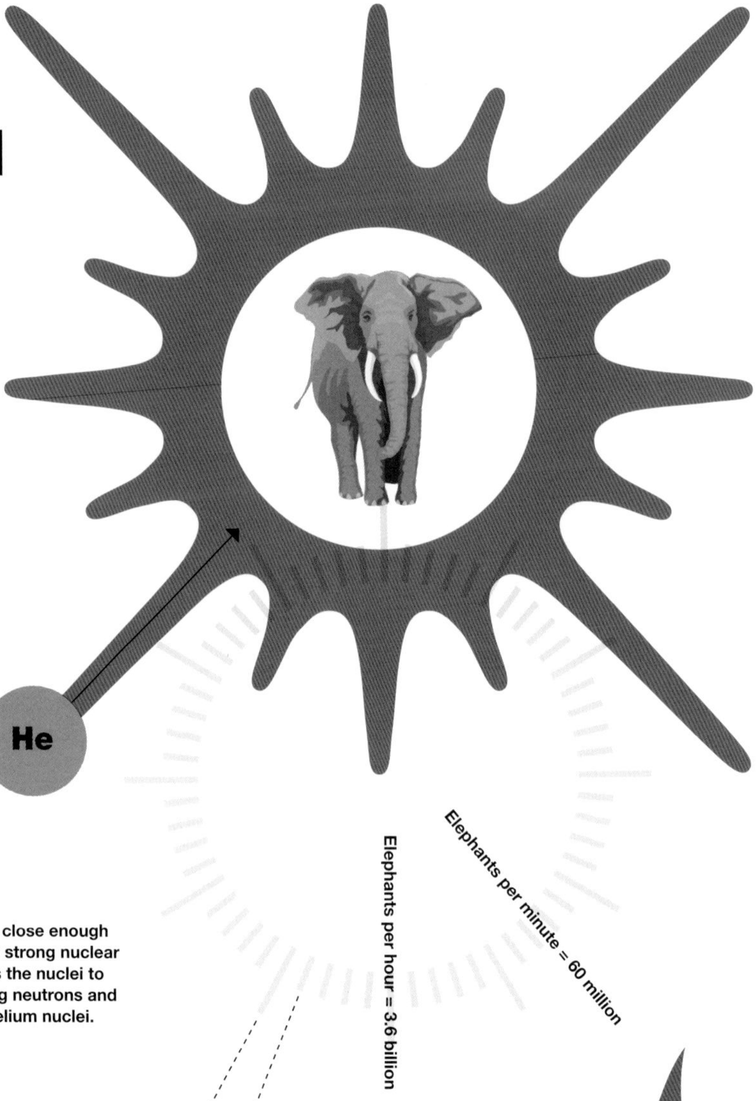

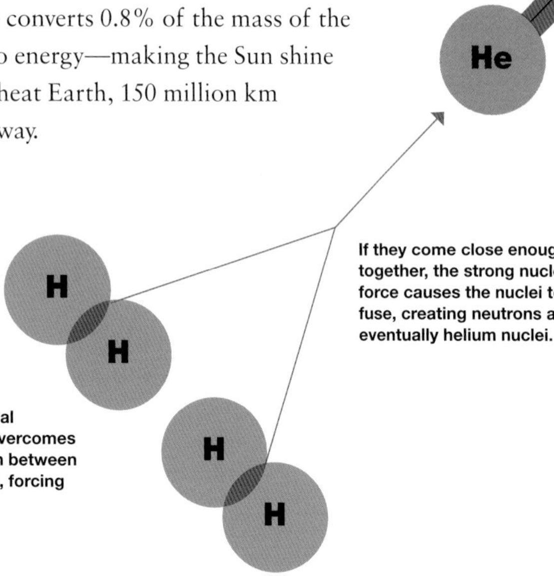

He

H

H

H

H

If they come close enough together, the strong nuclear force causes the nuclei to fuse, creating neutrons and eventually helium nuclei.

The enormous gravitational pressure inside the Sun overcomes the electrostatic repulsion between hydrogen nuclei (protons), forcing them together.

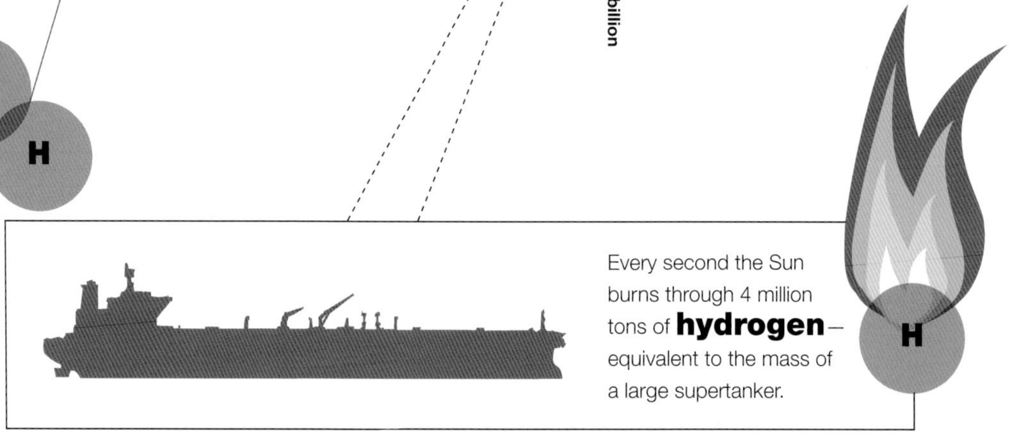

Elephants per minute = 60 million

Elephants per hour = 3.6 billion

Every second the Sun burns through 4 million tons of **hydrogen**—equivalent to the mass of a large supertanker.

H

X333,000

The Sun weighs 2,000 million billion trillion tons (2×10^{27} tons)—333,000 times more than Earth.

The Sun is so massive that even the vast amount of fuel it burns every second has little impact on its **overall mass**. The Sun loses just a 10-billion-billionth of a percent of its mass each second.

The Sun has lost just **0.1%** of its mass since it formed.

 Converting 1 kg (2.2 lb) of hydrogen into ~1 kg of helium releases a million times more energy than burning 1 kg of coal.

Every kilogram of hydrogen that the Sun consumes releases the same energy as a 1-megaton hydrogen bomb.

H

The Sun has a powerful **magnetic field** that twists and coils, and sometimes throws huge clouds of plasma (gas that has been heated to such high temperatures that it has been stripped of its electrons) into space. For instance, an explosion on October 28, 2010, caused a plasma twister 350,000 km (217,000 miles) high to erupt into space.

Solar flares are also caused by coils in the magnetic field. They can **release energy** equivalent to up to a billion megatonnes of TNT, flinging 100 billion tonnes of high-energy particles into space.

TNT

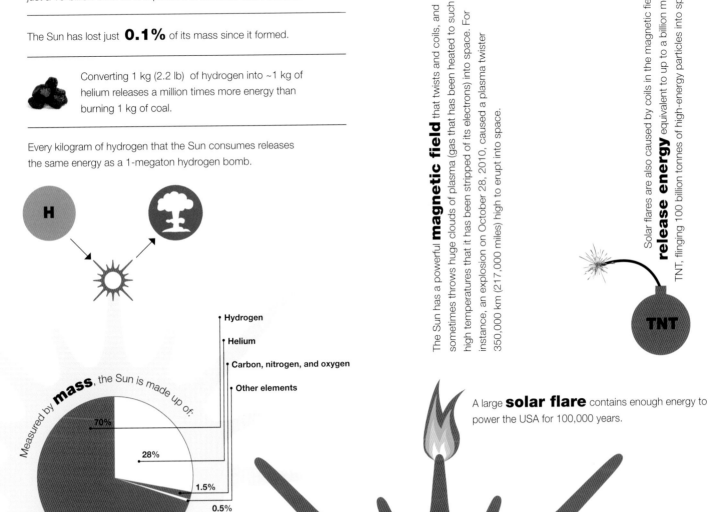

Measured by **mass**, the Sun is made up of:

- Hydrogen
- Helium
- Carbon, nitrogen, and oxygen
- Other elements

70%
28%
1.5%
0.5%

A large **solar flare** contains enough energy to power the USA for 100,000 years.

Traveling at the speed of light

▶ The speed of light is so huge that even a spaceship powered by hundreds of nuclear bombs—the fastest technology available—could only reach 5% of it.

The speed of light in a vacuum is the fastest possible speed in the universe. It is so vast that it makes the top speeds achieved by human technology seem minuscule, which in turn means that with current technologies interstellar travel is a pipe dream. The top speed attained by human space technology was over 240,000 km/h (150,000 mph), by the spacecraft *Helios 2*, which accelerated using the Sun's gravity like a slingshot.

An untested technology called nuclear pulse propulsion—which basically involves setting off a nuclear bomb underneath a spaceship once a second—might be able to accelerate a spacecraft to around 5% of the speed of light (54 million km/h, or 33.5 million mph).

In the 1960s U.S. government laboratories, under Project Orion, investigated a pulsed nuclear fission propulsion system. Small nuclear pulse units would be sequentially discharged from the aft end of the vehicle. A blast shield and shock absorber system would protect the crew and convert the shock waves into a continuous propulsive force.

Operating sequence of the Medusa propulsion system.

The bomb/pulse unit fires, beginning the process.

As the bomb's explosion pulse reaches the parachute canopy ...

... it pushes the canopy, accelerating it away from the bomb explosion as the spacecraft plays out the main tether with a winch, braking as it extends, and starts to accelerate the spacecraft ...

... before finally winching the tether back in.

Einstein's famous equation $E=mc^2$ gives the formula for the amount of energy, E, obtained by converting a mass, m, entirely into energy. Because the speed of light, c, is so enormous, the equation yields very high numbers. For instance, the theoretical energy content of 1 kg (2.2 lb) of matter would be enough to lift the entire population of the world into space.

 If the average person were converted entirely into energy, he would produce an explosion equivalent to **30 huge** H-bombs.

In fact it is almost **impossible** for 100% of mass to be converted into energy. Even in a spinning black hole, the most efficient mass-to-energy converter in the universe, only 43% of mass is converted into energy.

The *Voyager 1* spaceprobe used gravitational slingshotting to become the **fastest manmade object** now in existence. It is currently traveling at 62,000 km/h (38,600 mph).

It would take *Voyager 1* 73,000 years (2,500 generations) to get to Proxima Centauri.

Light travels at 1.079 billion km/h (671 million mph). It covers the distance between the Sun and Earth in just over eight minutes. In the course of a day, light travels 173 times further than this.

480 seconds

Helios 2, going at its maximum speed, would take 19,000 years to get there. If a community of people lived on board, they would pass through 600 generations on the trip.

4.22 years

85 years

The nearest star is Proxima Centauri. It takes **4.22** light years to travel from Earth to this star.

19,000 years

Nuclear pulses might be able to propel a spaceship to Proxima Centauri in just 85 years, although slowing down once you'd reached the star would be another matter.

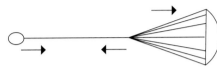

73,000 years

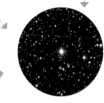

Proxima Centauri

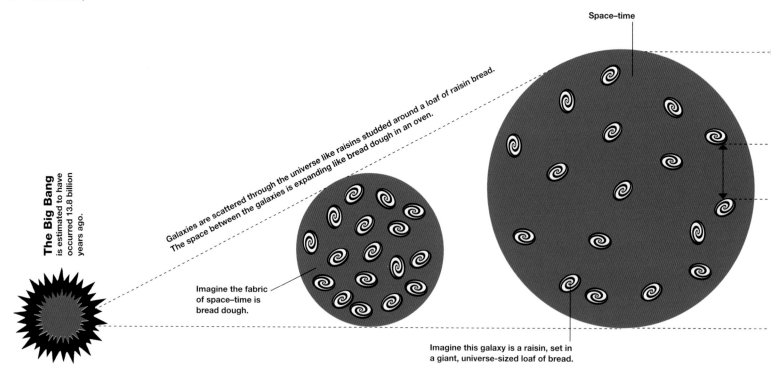

Space–time

The Big Bang is estimated to have occurred 13.8 billion years ago.

Galaxies are scattered through the universe like raisins studded around a loaf of raisin bread. The space between the galaxies is expanding like bread dough in an oven.

Imagine the fabric of space–time is bread dough.

Imagine this galaxy is a raisin, set in a giant, universe-sized loaf of bread.

The universe is like raisin bread

▶ Galaxies in an expanding universe are like raisins in bread that is put in the oven to rise: all the other raisins appear to be moving away at a speed proportional to their distance.

In 1929 American astronomer Edwin Hubble discovered that all of the galaxies he could observe were moving away from Earth, and each other, and they were doing so at speeds directly proportional to their distance away from one another. This is not the same as if they were all flying away from some central point, like shrapnel from an explosion. There is no center to the universe—the view is exactly the same wherever you are. The conclusion is that space itself is expanding.

This is analogous to raisin bread, where the raisins are equivalent to galaxies and the bread is the fabric of space–time. Imagine that the loaf is enormous, so that from the point of view of any individual raisin there is no surface or end to it. Now imagine that the raisin bread is put in the oven to rise. As the bread expands, all the raisins move away from one another. The greater the distance between any two raisins, the faster they move away from each other, and the same is true no matter which raisin you choose.

In the last **billion years** all of the space between clusters of galaxies has expanded by about

5%

The best current estimate for the **size of the universe** is that it is 93 billion light years across. But the universe is only 13.8 billion years old.

93 billion light years

All the detectable matter and energy in the universe only adds up to 5% of what scientists calculate must exist.

5%

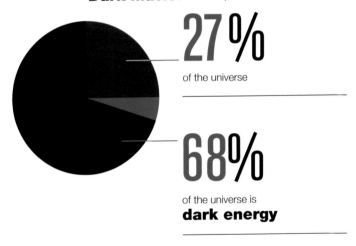

Dark matter makes up about

27%

of the universe

68%

of the universe is **dark energy**

The three possible shapes of our universe

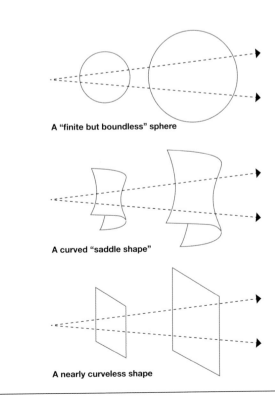

A "finite but boundless" sphere

A curved "saddle shape"

A nearly curveless shape

The universe does not add up. The force of gravity should be causing the universe's expansion to slow down. The fact that its expansion is accelerating shows that a mysterious force, known to astronomers as dark energy, is at work. Astronomers are also unable to account for all the mass that they know must exist in galaxies, so some sort of dark matter also exists.

The universe might be like a yo-yo

▶ If the universe is fated to end in a Big Crunch—the opposite of a Big Bang—it could lead to another Big Bang, so that the universe expands and contracts forever.

If the density of the universe is above a certain value, gravity will eventually overcome the outward expansion from the Big Bang, and it will collapse in on itself in a Big Crunch. This scenario is known as a closed universe, because it means that the geometry of the universe is spherical (a closed plane).

If the density is below this value, the universe will go on expanding forever. This is known as an open universe, and leads to a curved geometry like a saddle. If the universe has "critical density," gravity exactly matches the expanding force, and after an infinite amount of time the expansion will slow to zero. This is known as a flat universe.

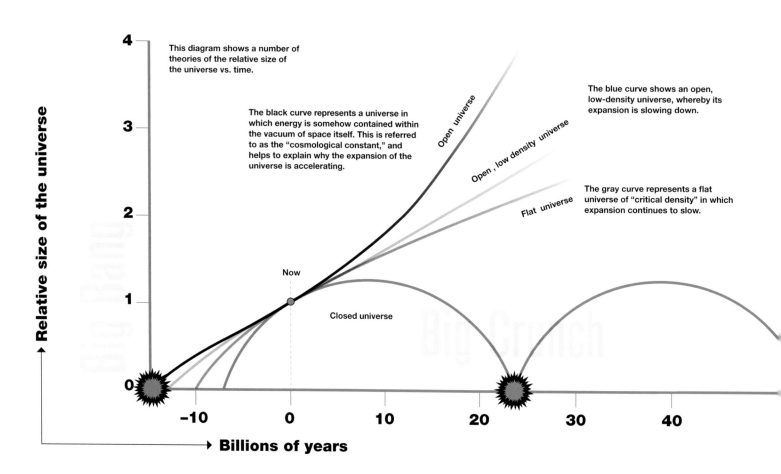

This diagram shows a number of theories of the relative size of the universe vs. time.

The black curve represents a universe in which energy is somehow contained within the vacuum of space itself. This is referred to as the "cosmological constant," and helps to explain why the expansion of the universe is accelerating.

The blue curve shows an open, low-density universe, whereby its expansion is slowing down.

The gray curve represents a flat universe of "critical density" in which expansion continues to slow.

Open universe

Open, low density universe

Flat universe

Now

Closed universe

Relative size of the universe

Billions of years

Imagine a spaceship launched from Earth. If it does not have the initial speed to escape Earth's gravity, it will fly in a big loop and come crunching back to Earth—as in a closed universe. If the spaceship has more than enough speed to escape Earth's gravity, it will continue speeding away forever—as in an open universe. If its speed exactly matches the pull of gravity, it will come to a stop after an infinite amount of time.

The critical density that separates an open universe from a closed universe is 10^{-26} g/cm^3. Your density is roughly 1 g/cm^3, but there are parts of the universe with a much lower density so the average might be close to the critical.

NASA's **WMAP** (Wilkinson Microwave Anisotropy Probe) measures Cosmic Microwave Background radiation (CMB) as a way of measuring its density. It tells us that our universe is flat, to within a 0.4% margin of error.

The Planck spacecraft was launched by the European Space Agency in 2009; it operated for four years.

The **CMB** is a remnant of the Big Bang. It is the coldest thing in nature, with one exception. The cosmic microwave background gives the universe an average temperature of 2.72778 K, (−270°C; −454°F).

The **coldest** point in space is the Boomerang Nebula at 1 K (−272.15°C; −457.87°F).

99% The CMB accounts for 99% of the radiation in the universe.

 If you take any sugar cube-sized volume of space it will contain **300 photons of CMB radiation**.

1% of the static on an untuned television set is accounted for by CMB.

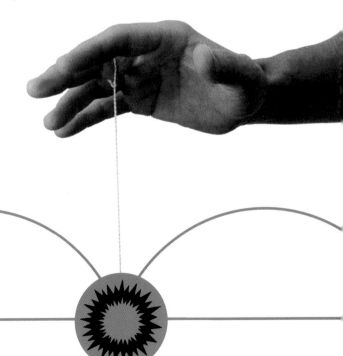

Looking for a firefly on the Moon

▶ The oldest and most distant galaxies are so faint that trying to spot them with a telescope is equivalent to trying to spot a firefly on the Moon.

In 2004 the Hubble Space Telescope took a picture known as the Ultra Deep Field (UDF), which shows 10,000 galaxies from the furthest—and therefore oldest—parts of the universe. Objects that formed when the universe was young are so distant that their light is incredibly faint; looking for them is therefore extremely challenging.

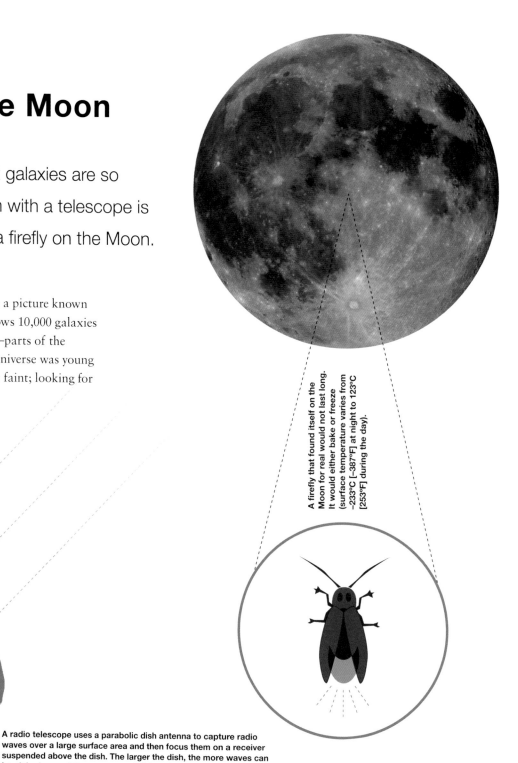

A firefly that found itself on the Moon for real would not last long. It would either bake or freeze (surface temperature varies from −233°C [−387°F] at night to 123°C [253°F] during the day).

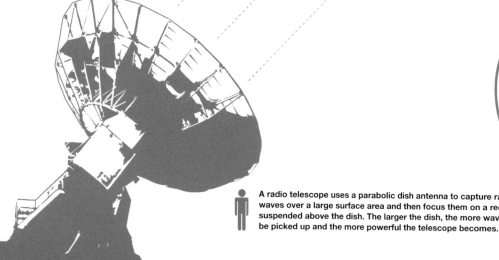

A radio telescope uses a parabolic dish antenna to capture radio waves over a large surface area and then focus them on a receiver suspended above the dish. The larger the dish, the more waves can be picked up and the more powerful the telescope becomes.

The whole sky contains
12.7 million times more area than the UDF.

Normally the Hubble telescope needs to capture millions and millions of **photons** per minute to take a picture. While taking the UDF picture, Hubble collected just one photon of light per minute.

To repeat this level of scrutiny over the entire night sky would take Hubble a million years of constant observation.

The Hubble Telescope has a **resolution** of

0.085

arcseconds.

This is equivalent to seeing the date on a U.S. quarter 2 km (1.25 miles) away.

Radio telescopes
detect radiation from the extremely low energy portion of the electromagnetic spectrum.

The VLBA radio telescope can detect motion roughly equivalent to the length of a baseball bat as seen from the Moon.

x12.7 million

The **UDF** is a small section of sky chosen for its relatively low density of bright stars in the "near-field" closer to Earth.

Ultra Deep Field

★6,000

About 6,000 stars in the night sky are visible to the naked eye from Earth, and only 3,000 from one person at any one spot.

★3,000

With binoculars, you can see around 50,000 stars.

★50,000

With a high-quality 16-inch telescope, you can see up to 100,000 galaxies.

x10,000 galaxies
There are 10,000 galaxies in the UDF.

The **faintest light** level that the human eye can perceive is of the 6th magnitude.

Hubble can detect objects at the **30th magnitude**

100,000

A supernova is like an exploding bottle of water

▶ A star going supernova blows off its outer layers of gas like water exploding out of the top of a bottle of water that is banged on a table.

A star of mass greater than eight times the mass of the Sun ends its life as a supernova. When it runs out of fuel for the fusion reaction that sustains it, it no longer produces the outward radiant energy pressure that stops it from collapsing. All that is left is a huge ball of matter, which implodes under the force of its own gravity. The violent shockwave of this implosion compacts the matter of the star into ultra-dense neutrons (see pp. 132–3), the densest form of matter possible. When the implosion's shockwave runs up against this ultimate density, it rebounds outwards with massive force, blowing off the outer gaseous shell of the star and scattering heavy elements into space in the form of a nebula. The explosion is similar to what happens if you slam the bottom of an open bottle of water against a table. The sudden deceleration of the water creates a shockwave that travels down the bottle and rebounds upward, blowing most of the water out of the top.

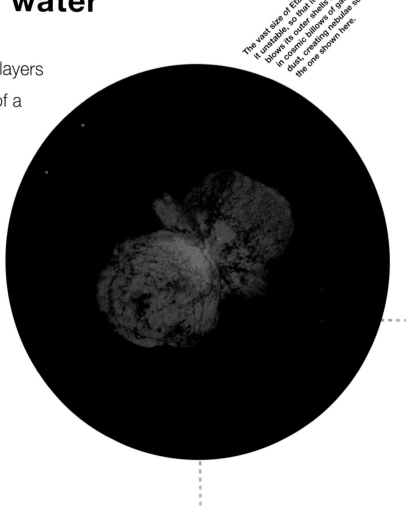

The vast size of Eta Carinae makes it unstable, so that it periodically blows its outer shells into space in cosmic billows of gas and dust, creating nebulae such as the one shown here.

x10²⁷

The **energy given off** by a supernova is equivalent to 1,000 trillion trillion hydrogen bombs going off at once.

For a few months, a supernova shines more brightly than an entire galaxy, as bright as

100 **billion stars**

x100,000trillion

In 2005 NASA detected a collision between two neutron stars traveling at over 32,000 km/s (20,000 miles/s), which emitted as much light as 100,000 trillion suns.

A supernova star collapses to a hundred-thousandth of its diameter. Our Sun is too small to go supernova, but if it did it would collapse to give a neutron star just 15 km (9.5 miles) across.

$$\frac{1}{100,000}$$

100 x heavier

400 x wider

4,000,000 x brighter

The most massive star known is **R136a1** in the Tarantula Nebula. It is 315 times heavier than the Sun, 29 times wider, and nearly 9 million times brighter.

x1

In a typical galaxy, there will be one supernova **every 50 years**

In 1987 this supernova, SN 1987a, became the brightest in recent history, but its most notable features—the rings surrounding the remains of the star—remain mysterious and unexplained.

The shockwaves generated by a supernova radiate outward at

35 **million**km/h

(20 million mph).

The largest known star in the universe is UY Scuti, a red hypergiant

about 950 light years away in the constellation Scutum. It may be more than 1708 times larger than the Sun—if the Sun were this large it would extend almost to the orbit of Saturn. Light takes more than two hours to get from one side of the star to the other.

Section Five

▶ We spend our whole lives on Planet Earth, yet there is so much we don't understand about it. This section will illustrate the awesome power of natural forces such as earthquakes, volcanoes, hurricanes, and tsunamis, and help to explain the geological formation and composition of Earth, and the behavior of its atmosphere and weather.

Earth Science

A mighty wind

▶ In a single day, a major hurricane generates enough electricity to power the whole of Britain for a year.

A tropical cyclone, or hurricane, is created when air picks up moisture and heat from the tropical ocean. As the warm, wet air rises, its water vapor condenses into water droplets. When water vapor condenses it releases energy, known as the latent heat of condensation. Some of this energy is converted into mechanical energy, aka wind, and this wind then blows across the ocean picking up more heat and vapor, in a positive feedback loop. If the conditions are right, such a system can build up huge amounts of energy, making hurricanes enormously powerful and destructive.

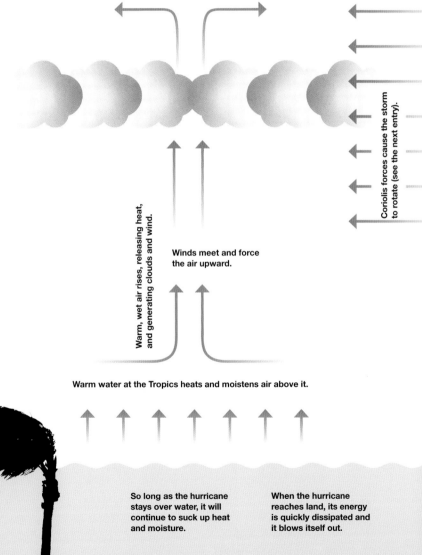

Coriolis forces cause the storm to rotate (see the next entry).

Warm, wet air rises, releasing heat, and generating clouds and wind.

Winds meet and force the air upward.

Warm water at the Tropics heats and moistens air above it.

So long as the hurricane stays over water, it will continue to suck up heat and moisture.

When the hurricane reaches land, its energy is quickly dissipated and it blows itself out.

**x400**

In a single day, a large hurricane can **generate energy** equivalent to 400 20-megatonne H-bombs, or

8,000 megatonnes

(8 billion tonnes). The destructive power of the world's nuclear arsenal is equivalent to around 7,000 megatonnes.

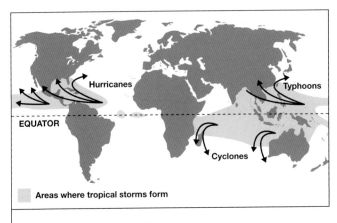

Hurricanes

Typhoons

EQUATOR

Cyclones

Areas where tropical storms form

Tropical storms are called by different names in different parts of the world. In the Atlantic and around North, Central, and South America they are called hurricanes. In the Far East they are known as typhoons, and in the Indian Ocean and Australia they are known as cyclones.

The **energy** released by a single hurricane could power the entire USA for six months, or provide a whole year of power for a country like Britain or France.

4

days

Even a normal thunderstorm generates power equivalent to the electricity consumption of the entire USA for

At any moment there are **1,800** thunderstorms in progress around the world—that's around 40,000 storms a day.

Every second

Earth is struck by 100 lightning bolts.

100

Clouds are surprisingly insubstantial. A typical cumulus cloud contains just enough water to fill a small bathtub.

If you walk through 100 m of **fog** you will pick up about 8 ml (¼ fl oz) of water—barely a mouthful.

Throwing a ball off a merry-go-round

▶ The Coriolis effect is like standing in the middle of a merry-go-round
and throwing a ball to someone riding a horse at the edge.

The Coriolis effect causes anything traveling from one latitude to another to veer off from a straight line. When air masses move from north or south toward the equator, they are deflected to the west. This causes cyclones to spin counterclockwise in the northern hemisphere and clockwise in the southern hemisphere. The Coriolis effect is a consequence of Earth's near spherical shape, which means that a point at the equator is traveling faster than points at higher latitudes, just as the perimeter of a merry-go-round is traveling faster than the center. If you stand in the middle of a merry-go-round and throw a ball to someone riding a horse near the edge, the ball will have no lateral velocity—it will go in a straight line. But to the person on the horse who does have lateral velocity, the ball will appear to curve away from him because by the time the ball reaches where he was when it was thrown, his horse has moved on.

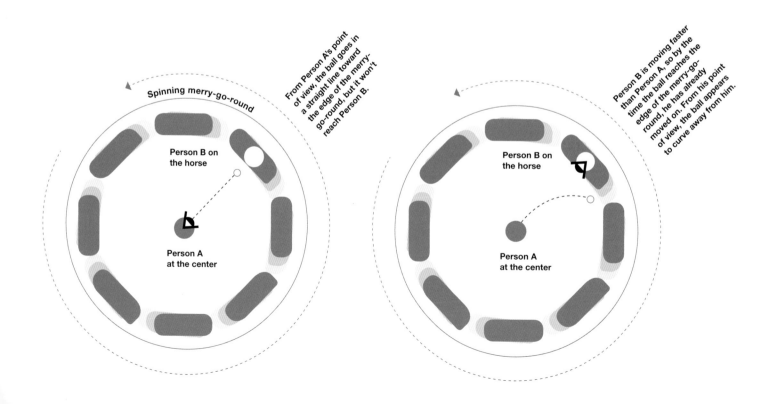

From Person A's point of view, the ball goes in a straight line toward the edge of the merry-go-round, but it won't reach Person B.

Spinning merry-go-round

Person B on the horse

Person A at the center

Person B is moving faster than Person A, so by the time the ball reaches the edge of the merry-go-round, he has already moved on. From his point of view, the ball appears to curve away from him.

Person B on the horse

Person A at the center

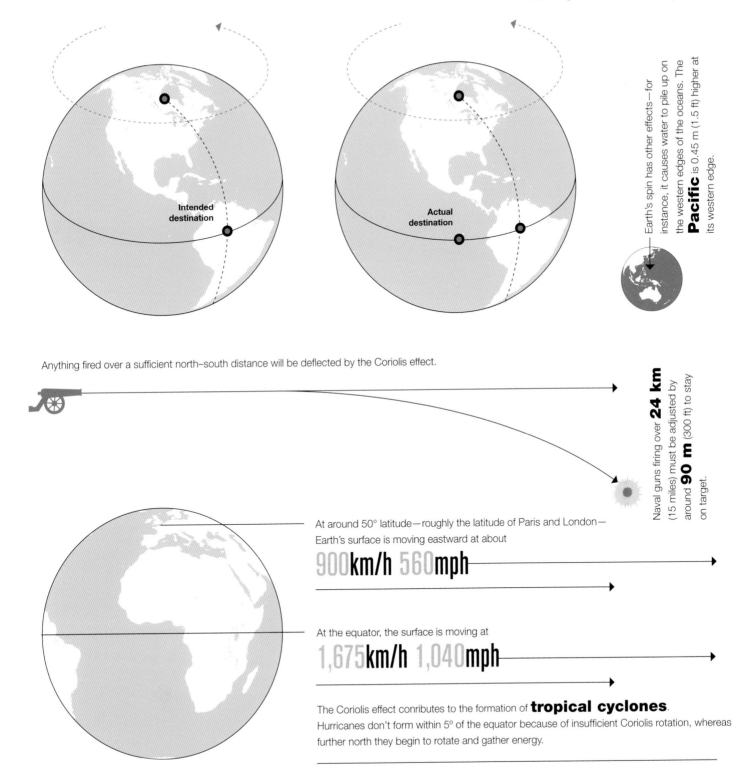

Intended destination

Actual destination

Earth's spin has other effects—for instance, it causes water to pile up on the western edges of the oceans. The **Pacific** is 0.45 m (1.5 ft) higher at its western edge.

Anything fired over a sufficient north–south distance will be deflected by the Coriolis effect.

Naval guns firing over **24 km** (15 miles) must be adjusted by around **90 m** (300 ft) to stay on target.

At around 50° latitude—roughly the latitude of Paris and London—Earth's surface is moving eastward at about

900km/h 560mph

At the equator, the surface is moving at

1,675km/h 1,040mph

The Coriolis effect conributes to the formation of **tropical cyclones**.
Hurricanes don't form within 5° of the equator because of insufficient Coriolis rotation, whereas further north they begin to rotate and gather energy.

Landfilling the Grand Canyon

If the Grand Canyon were a landfill site, it would take nearly 14,000 years to fill it with trash.

Geological processes operate on vast scales over vast time periods. Few places on Earth illustrate this quite so graphically as the Grand Canyon, which runs for 446 km (277 miles) through the state of Arizona. Formed by the Colorado River, the Canyon is up to a 1.6 km (1 mile) deep and 16 km (10 miles) wide. According to the U.S. National Parks Service, the volume of the Grand Canyon is 4.17 trillion cubic meters (5.45 trillion cubic yards). It would take three Grand Canyons to hold Lake Superior, the largest freshwater lake in the world (by surface area) and five and a half to hold Lake Baikal (the largest by volume).

According to the Environmental Protection Agency, the U.S. discarded **169 million tonnes** of municipal waste to landfills in 2014, the most recent year for which figures are available.

The **density** of landfill varies, but the average is about 450 kg (1,000 lb)/m³. This means that at 2014 rates the annual volume of landfill produced by the USA is 300 million cubic meters (10.6 billion cubic feet).

At this rate, it would take 13,900 years to fill the Grand Canyon with garbage, by which time it would contain enough trash to cover Earth to a depth of 1 cm (⅖ in).

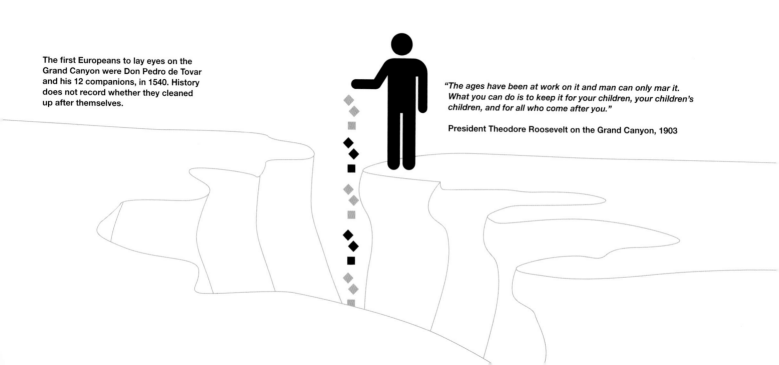

The first Europeans to lay eyes on the Grand Canyon were Don Pedro de Tovar and his 12 companions, in 1540. History does not record whether they cleaned up after themselves.

"The ages have been at work on it and man can only mar it. What you can do is to keep it for your children, your children's children, and for all who come after you."

President Theodore Roosevelt on the Grand Canyon, 1903

The **Colorado River** has carved out the Canyon as the land has risen over the last 6 million years, but in doing so it has exposed some of the oldest rock strata on Earth, going back 2 billion years, spanning four of the major geological eras of the planet, from the late Precambrian, through the Palaeozoic and Mesozoic to the Cenozoic.

2 billion years

Earth's Grand Canyon is dwarfed by its Martian equivalent, the **Valles Marineris**, which is up to 200 km (125 miles) wide and 8 km (5 miles) deep at points.

It stretches around almost a quarter of Mars' equator— over 4,000 km (2,500 miles)—and would stretch across the entire USA if moved to Earth.

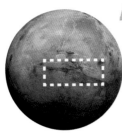

4,000 km 2,500 miles long

7 km 4.3 miles deep

200 km 125 miles wide

9 km deep

Deeper still is **Melas Chasma**, the deepest canyon in the Solar System, which reaches 9 km (5.6 miles) below the plateau on Mars in which it sits.

Although the Grand Canyon is rich in evidence of prehistoric life, no fossil bones have ever been recovered from its rocks.

Perhaps this is not surprising—only around **15%** of rocks are capable of preserving fossils.

Human fossils are particularly rare. All the hominid bones ever discovered could fit in the back of a pickup truck. Out of the billions of individuals who have ever lived, only 5,000 have left remains. The number of known *Homo erectus* individuals would not even fill a school bus.

Even arid soils, like those that make up much of the Grand Canyon National Park, contain **billions of bacteria**. A handful of soil from the woods that cover the Park's higher elevations might contain up to

10,000,000,000 bacteria
1,000,000 yeast cells
200,000 molds
10,000 protozoans

Even more bacteria are found deep below the surface of Earth. It is estimated that the deep strata of the planet may harbor over **100 trillion tonnes** of bacteria—enough to cover the surface of Earth to a depth of 15 m (50 ft), the height of a four-story building.

Drinking the Pacific

▶ If the Pacific Ocean were drinking water, it would take you 960,000 trillion years to drink it.

Water is the defining feature of Earth—it is often suggested that the planet is misnamed. Most of the planet's surface is covered in water; in fact most of it is under more than 1.6 km (1 mile) of water. The vast majority of the planet's water is in the oceans, and by far the biggest of these is the Pacific.

Somebody drinking a healthy eight glasses of water a day would take 350 billion billion days, or 960,000 trillion years, to drink the Pacific. This is 70 million times the age of the universe.

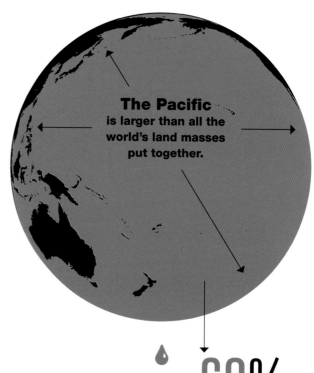

The Pacific is larger than all the world's land masses put together.

60%

of Earth's surface is under more than **1.6 km** (1 mile) of water. If Earth were smooth it would be covered with water to a depth of 4 km (2.5 miles).

There are 700 million cubic km (700,000 trillion cubic meters) in the Pacific Ocean— equivalent to 2,800 million trillion glasses of water. That's

700milliontrillionliters

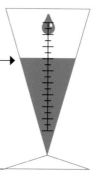

There are 1.3 billion cubic km of water on Earth, and there will never be any more. Most of this water had arrived by 3.8 billion years ago, so any glass of water you drink is

3.8billion years old

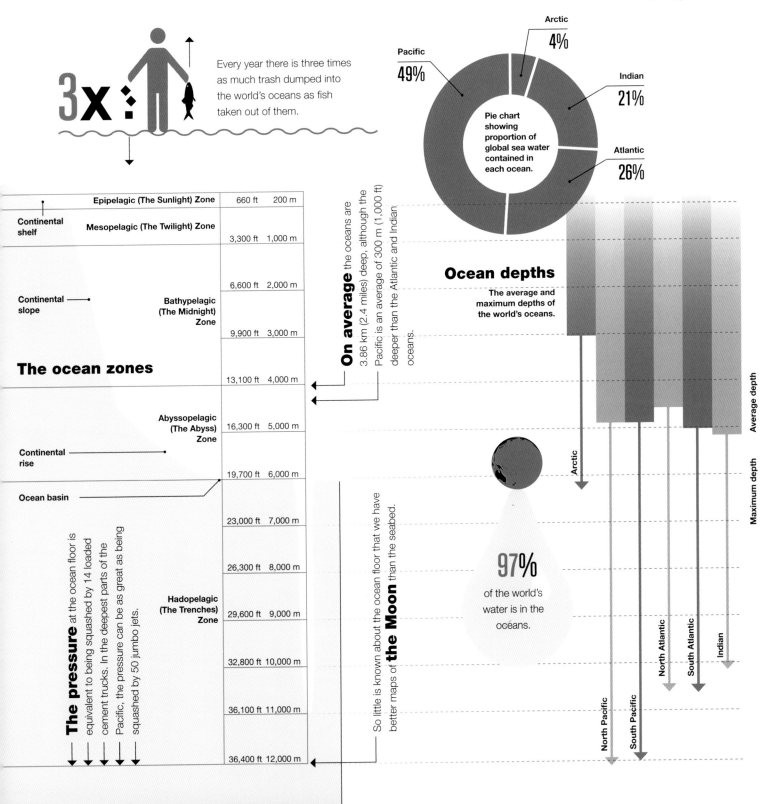

3x Every year there is three times as much trash dumped into the world's oceans as fish taken out of them.

Pie chart showing proportion of global sea water contained in each ocean.

Pacific
49%

Arctic
4%

Indian
21%

Atlantic
26%

Epipelagic (The Sunlight) Zone	660 ft	200 m
Mesopelagic (The Twilight) Zone		
	3,300 ft	1,000 m
	6,600 ft	2,000 m
Bathypelagic (The Midnight) Zone		
	9,900 ft	3,000 m
	13,100 ft	4,000 m
Abyssopelagic (The Abyss) Zone		
	16,300 ft	5,000 m
	19,700 ft	6,000 m
	23,000 ft	7,000 m
	26,300 ft	8,000 m
Hadopelagic (The Trenches) Zone		
	29,600 ft	9,000 m
	32,800 ft	10,000 m
	36,100 ft	11,000 m
	36,400 ft	12,000 m

Continental shelf

Continental slope

The ocean zones

Continental rise

Ocean basin

On average the oceans are 3.86 km (2.4 miles) deep, although the Pacific is an average of 300 m (1,000 ft) deeper than the Atlantic and Indian oceans.

The pressure at the ocean floor is equivalent to being squashed by 14 loaded cement trucks. In the deepest parts of the Pacific, the pressure can be as great as being squashed by 50 jumbo jets.

So little is known about the ocean floor that we have better maps of **the Moon** than the seabed.

Ocean depths
The average and maximum depths of the world's oceans.

Average depth

Maximum depth

Arctic

97% of the world's water is in the oceans.

North Pacific

South Pacific

North Atlantic

South Atlantic

Indian

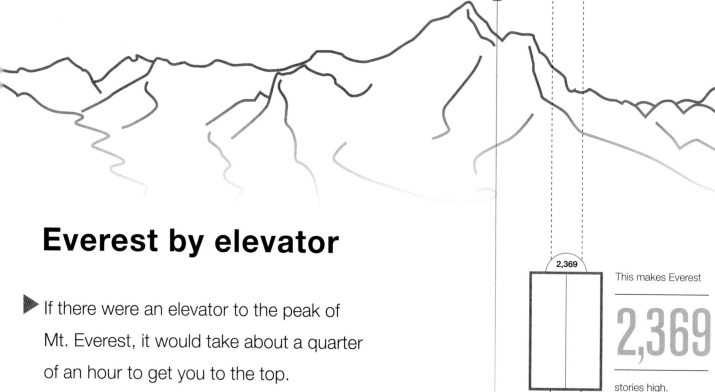

Everest by elevator

▶ If there were an elevator to the peak of Mt. Everest, it would take about a quarter of an hour to get you to the top.

The Himalayas are proof of the powerful forces at work in Earth's crust–pushed miles into the sky by the Indo-Australian tectonic plate crashing into the Eurasian plate, it is the greatest mountain range on Earth yet has taken just 55 million years to form. These immense forces are still at work: as the Indian plate drives under the Tibetan plateau, the Himalayas are effectively advancing across India.

Mt. Everest is 8,848 m (29,035 ft) **high**—that's

8.85km 5.5miles

2,369

This makes Everest

2,369

stories high.

It would take 20 Empire State Buildings stacked on top of each other to reach **the peak**.

x20

Although several companies are currently working on elevators that will travel at **64 km/h** (40 mph), the elevators in the world's tallest building, the Burj Khalifa (see pp. 204–5), travel at 35 km/h (22 mph). At this speed it would take a 15-minute elevator ride to get from sea level to the top of Everest.

1 centimeter

The Himalayas **continue to rise** by more than 1 cm (0.4 in) a year.

Around 60 million years ago the **Indo-Australian** plate was moving at around 15 cm (6 in) a year. Even today it plows northward at around 5 cm (2 in) a year. In 10 million years' time India will have advanced another 180 km (110 miles) into (or, rather, under) Tibet.

Over the next 10 million years, Nepal will cease to exist, crushed out of existence by the power of tectonics.

The Himalayan mountain range is so large that it is visible from space.

Mountain ranges like the Himalayas are ground down by the glaciers and ice sheets they develop. Technically, we are still in a period of **global cooling**, characterized by repeated ice ages—17 of them since the first humans evolved.

30%

During the last Ice Age, 30% of the land surface of Earth was under ice. At its height, the ice sheet over the northern land masses advanced at 120 m (393 ft) a year.

If not disrupted by **manmade global warming**, there would be another 50 or so Ice Ages, each lasting about 100,000 years, before the next period of natural global warming.

Tectonic plate boundaries

The boundary where two plates meet is a zone of tectonic activity. Plate boundaries can take three forms—convergent, divergent, and transform—depending on whether they are moving toward, away from, or parallel to each other.

Convergent: Where two plates are crashing together (as with the Indian and Eurasian plate), one is subducted beneath the other, which experiences uplift.

Divergent: Where new rock is being created, plates are forced apart, as at the North Atlantic Ridge or Rift Valley.

Transform: Where two plates slide past each other, as along the San Andreas Fault in California, they rarely do so smoothly, moving in fits and starts, causing earthquakes.

Measured from base to peak, the Hawaiian seamount **Mauna Kea** is the tallest mountain on Earth at 10.2 km (6.34 miles).

The **deepest** point on Earth is the Challenger Deep at the bottom of the Marianas Trench in the Pacific, which is just over 11 km (6½ miles) deep—the height of Mt. Everest with four CN Towers stacked on top.

Mauna Kea, Earth

Even this is less than half the height of the **highest mountain in the Solar System**, Rheasilvia, on the asteroid Vesta. It towers 22 kilometres (13.8 miles) over the bottom of a crater.

Rheasilvia, Vesta

The Challenger Deep

Like an H-bomb in San Francisco

▶ So much energy was released in the San Francisco earthquake of 1906 that it was as if one of the largest H-bombs ever created had been set off under the city.

When tectonic plates slide or grind past or under one another, they do not always do so smoothly. Often they will travel in fits and starts, with enormous tension building up over time until it is released in a sudden jolt—an earthquake. The Richter Scale was devised in 1935 by Charles F. Richter of the California Institute of Technology as a mathematical device to compare the size of earthquakes. It is a logarithmic scale (see pp. 88–9), which measures the amount of movement of the ground caused by the quake (its amplitude). An increase of 1 on the Richter Scale is therefore equivalent to a tenfold increase in amplitude. The amount of energy released increases even more—by 31 times—with every step up the scale. Although higher magnitude quakes are more energetic, the hazard a quake poses is as much to do with the depth of the focus (the epicenter of the quake) and the nature of the surrounding rock.

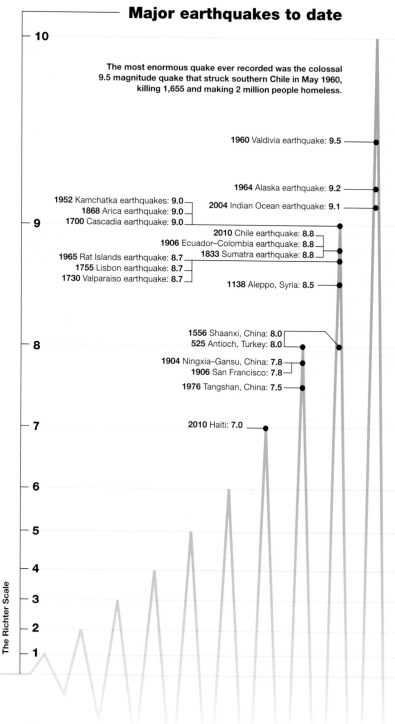

Major earthquakes to date

The most enormous quake ever recorded was the colossal 9.5 magnitude quake that struck southern Chile in May 1960, killing 1,655 and making 2 million people homeless.

1960 Valdivia earthquake: **9.5**

1964 Alaska earthquake: **9.2**

1952 Kamchatka earthquakes: **9.0**
1868 Arica earthquake: **9.0**
1700 Cascadia earthquake: **9.0**

2004 Indian Ocean earthquake: **9.1**

2010 Chile earthquake: **8.8**
1906 Ecuador–Colombia earthquake: **8.8**
1833 Sumatra earthquake: **8.8**

1965 Rat Islands earthquake: **8.7**
1755 Lisbon earthquake: **8.7**
1730 Valparaiso earthquake: **8.7**

1138 Aleppo, Syria: **8.5**

1556 Shaanxi, China: **8.0**
525 Antioch, Turkey: **8.0**

1904 Ningxia–Gansu, China: **7.8**
1906 San Francisco: **7.8**

1976 Tangshan, China: **7.5**

2010 Haiti: **7.0**

The Richter Scale

A seismograph trace is recorded by a pen attached to an armature that is incredibly sensitive to vibrations, and therefore moves up and down in response to even the slightest tremors.

Energy release equivalents

The Richter Scale for measuring earthquake magnitude is a logarithmic scale, which means that each increment is an order of magnitude. Thus a score of 10 on the Richter Scale is not simply 10 times greater than 1, but 10^{10} (10 billion) times greater. There is no theoretical limit to the Richter Scale, but earthquakes measuring higher than 10 would be literally Earth-shattering events.

Krakatoa eruption 1883

6-megatonne nuclear bomb

Mt. St. Helens eruption 1980

Hiroshima atomic bomb 1945

Average tornado

Large lightning bolt

Oklahoma City bomb 1995

Despite the enormous forces involved, **tectonic plates** generally move extremely slowly—as fast as a fingernail grows. Tectonics are driven by the flow of partially liquid mantle rock beneath the plates, which crawls along at roughly one ten-thousandth the speed of the hour hand on a clock.

A **magnitude 8** earthquake releases as much energy as a 6-megatonne nuclear bomb.

Around four times a year there is a quake of magnitude **7.4** or more—powerful enough to bring down most buildings.

Above **5.5**, quakes begin to damage houses; above **6.2** they become really dangerous.

Quakes of **4.3–4.8** will rattle the windows—these occur nearly 5,000 times a year.

Earthquakes of around magnitude **4.0** are faintly detectable indoors.

Earthquakes of less than magnitude **3.4** are not noticeable to people, but occur over 800,000 times a year.

Earthquakes of **2.0** or less are known as microearthquakes and are only recorded by local seismometers.

A magnitude **1.0** earthquake is almost undetectable, and releases energy equivalent to blowing up 170 g (6 oz) of TNT.

Volcanoes—nature's weapons of mass destruction

▶ If the magma lake below Yellowstone National Park were a pile of dynamite, it would be as big as an English county and as high as the troposphere.

Normal volcanoes may be terrifying and destructive, but they pale by comparison with supervolcanoes—geological events that threaten the survival of mankind, and even of life on Earth. The last supervolcano was the eruption of the Taupo Volcano in New Zealand, which exploded 26,500 years ago. The magnitude of a volcanic eruption is measured with the Volcanic Explosivity Index (VEI), a logarithmic scale like the Richter.

VEI	Ejecta volume	Classification	Description	Plume	Typical frequency	Example	Occurrences in the last 10,000 years
8	> 1,000 km³	Ultra-plinian	Mega-colossal	> 25 km	≥ 10,000 yrs	Taupo (26,500 BP)	0
7	> 100 km³	Plinian/Ultra-plinian	Super-colossal	> 25 km	≥ 1,000 yrs	Tambora (1815)	5 (+2 suspected)
6	> 10 km³	Plinian/Ultra-plinian	Colossal	> 25 km	≥ 100 yrs	Krakatoa (1883)	51
5	> 1 km³	Plinian	Paroxysmal	> 25 km	≥ 50 yrs	Mt. St. Helens (1980)	166
4	> 0.1 km³	Peléan/Plinian	Cataclysmic	10–25 km	≥ 10 yrs	Eyjafjallajökull (2010)	421
3	> 10,000,000 m³	Vulcanian/Peléan	Severe	3–15 km	Yearly	Cordón Caulle (1921)	868
2	> 1,000,000 m³	Strombolian/Vulcanian	Explosive	1–5 km	Weekly	Galeras (1993)	3,477
1	> 10,000 m³	Hawaiian/Strombolian	Gentle	100–1,000 m	Daily	Stromboli	many
0	< 10,000 m³	Hawaiian	Non-explosive	< 100 m	Constant	Mauna Loa	many

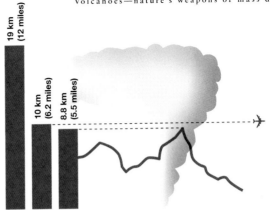

The most powerful eruption of historic times was Mt. Tambora in 1815, a VEI 7 event. Toba was a VEI 8 event, which is thought to have killed almost every human on the planet. Yellowstone is the site of previous super-volcanic eruptions, and is sitting on top of a magma reservoir 70 km (43 miles) long and 13 km (8 miles) deep.

Mount St. Helens erupted on May 18, 1980, producing a cloud of debris 19 km (12 miles) high—more than twice the cruising altitude of a jetliner or than the height of Mt. Everest.

540 million

tonnes of ash from Mt. St. Helens fell over an area of more than 60,000 square kilometers (22,000 square miles)—an area slightly larger than Egypt—in just

9 hours

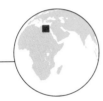

A supervolcanic eruption at **Yellowstone** would have an effect equivalent to the impact of a 1.5 km- or 1 mile-diameter asteroid. The explosion would blast an 80 km (50 mile) wide hole in Earth's crust. An area 1,000 km² would be obliterated in minutes. Ash from the eruption would blanket three-quarters of the USA to a depth of many centimeters.

The aerosol generated by such an eruption would blanket Earth for years, potentially blocking 99% of sunlight and causing a volcanic winter with temperature drops of 5–10°C, possibly as much as 15°C around the Tropics.

The **Toba** eruption (75,000 years ago) blasted between 3,000 and 6,000 km³ of material into the atmosphere, and enough sulfur to create 5 billion tonnes of sulfur aerosol in the stratosphere. This is up to 50 times as much material as ejected by Tambora.

After the Toba eruption, it took 20,000 years for the human population of Earth to recover beyond a few thousand individuals.

Tambora blasted material into the sky at a rate of 300,000 tonnes/second. Toba was ten times more powerful.

Even larger than supervolcanoes are flood basalt events, where massive flows of lava are released over centuries or more. These can release hundreds of thousands of cubic kilometers of material, covering up to a million square kilometers.

3

The Siberian Traps flood basalt event of 250 million years ago released lava for a million years—enough to cover the entire planet to a depth of

meters (10 feet)

Waves of death

▶ A tsunami is like the wave created if you shove yourself backward in the bath— unlike normal waves, which are formed by wind action, a tsunami is generated by rapid displacement of water, just like a big bath splosh.

If you sit in a bathtub full of water and suddenly extend your legs, pushing yourself off one end of the bath and shunting yourself backward, you will displace water so violently that most of it will slop out of the bath in a big wave. This is analogous to what happens when a tsunami is triggered by an earthquake. Sudden uplift of the sea floor displaces all the water above it, and although the actual degree of earth movement may be relatively small, the colossal mass of water displaced generates a wave of immense energy. In deep water this wave passes by without disturbing the surface by more than a few meters, but when it reaches shallow coastal areas it piles up into a monstrous wave or tsunami.

Tsunami waves have quite large amplitudes (meaning they are high), but what really does the damage is the long wavelength. Until the whole wavelength of the tsunami has crossed the shore, the wave keeps coming, a phenomenon graphically demonstrated during the Boxing Day tsunami of 2004.

A mega-tsunami is a particularly massive displacement of water created by a huge landslide. Such an event might generate a huge wave that crosses an entire ocean, like dropping a brick into one end of a shallow trough or trench of water and causing an overflow at the other end.

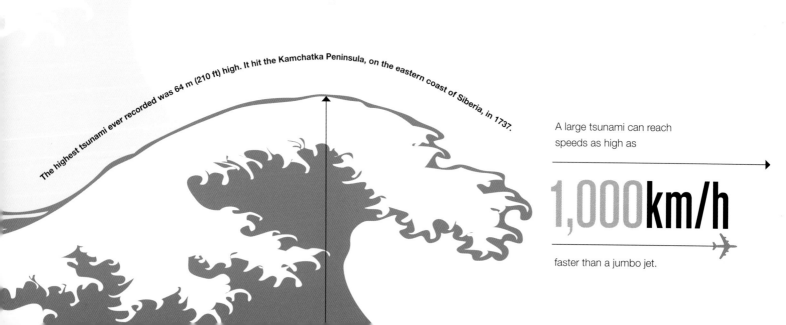

The highest tsunami ever recorded was 64 m (210 ft) high. It hit the Kamchatka Peninsula, on the eastern coast of Siberia, in 1737.

A large tsunami can reach speeds as high as

1,000km/h

faster than a jumbo jet.

One of the most likely spots for a mega-tsunami to form is the **Cumbre Vieja** volcano on La Palma in the Canaries. Its western flank was slightly displaced during an earthquake in 1949 and some experts believe it could collapse given further tremors. The weight of rock amounts to

500 billion tonnes

According to Professor Bill McGuire, the resulting mega-tsunami would be 900 m (nearly 3,000 ft) high at its **epicenter**, and 100 m (330 ft) high when it hits the other Canary Islands.

It would **radiate** across the Atlantic at near supersonic speeds, hitting the coast of Africa after one hour, and Spain, Britain, and Ireland five to seven hours later, at which point it would still be 7 m (23 ft) high.

After 9–12 hours, it would reach the eastern seaboard of the **USA**, where estuaries and bays could funnel it into 50 m (165 ft) waves.

With a **wavelength** hundreds of kilometers long, the mega-tsunami would keep coming for 15 minutes, advancing far inland, before flooding back the other way for a further 15 minutes.

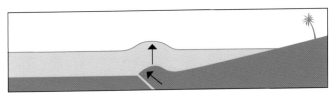

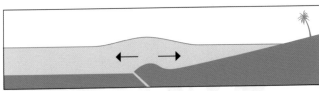

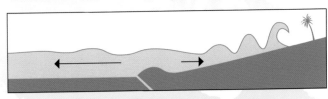

Ocean floor uplift along a fault line lifts the column of water above it. This generates a huge dome of water, which then subsides and ripples outward in all directions. When it hits shallow water, the wave rears up to create a tsunami.

A few times a year, normal wave-generating processes, such as the action of wind, produce rogue or freak waves that are over **30 m (100 ft) high** —the height of a 12-story building.

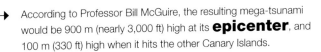
7 m

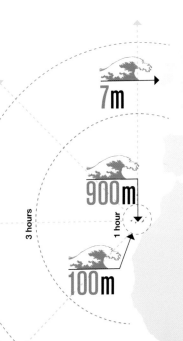
50 m

900 m

100 m

6 hours

3 hours

1 hour

It came from outer space

▶ Every year the mass of Earth increases by the equivalent of a cruise liner, thanks to a constant rain of space dust.

When Earth was young, the Solar System was a much more dangerous place, with massive asteroids and even planetoids battering the newborn planet. Today it is far safer but space is still full of billions of pieces of rock and debris, many of them close to Earth—these are known as Near Earth Objects (NEO). Tiny dust grains known as cosmic spherules rain down constantly on the planet, and larger rocks regularly burn up in our atmosphere. Fortunately the risk of impact decreases as the potential size of impactor increases.

To help quantify the true risk posed by a NEO, Richard Binzel of the Massachusetts Institute of Technology (MIT) devised the Torino scale, which takes into account both the probability of an impact and the scale of damage that it would cause. A Torino score of 0 means that either an impact has virtually no chance of happening, or that it will not cause any significant damage. A Torino score of 10 refers to an impact that is 100% certain to happen, and that will devastate life on Earth.

1 million small meteoroids

have hit the Earth during the past 24 hours.

30,000 tonnes

of space dust are collected each **year** by Earth.

There are about a million **NEOs** that could do substantial damage if one hit Earth.

Even more dangerous are **ECAs** (Earth Crossing Asteroids). It is estimated that there are 100,000 ECAs 100 m+ in diameter, 20,000 500 m+ and, according to NASA's Jet Propulsion Lab, there are between 500 and 1,000 ECAs bigger than 1 km (0.6 miles) in diameter.

More than 300 of these have been observed and are known not to pose a **danger** for at least several hundred years. In fact, out of all the known ECAs, only 13 are believed to have any chance of crossing our orbit before 2100.

x190

There are 190 known impact craters on the planet. Of the 25 massive **extinctions** known from the fossil records, seven of them have been linked to impacts.

Meteor Crater in Arizona, where the arid conditions have preservedit from erosion and kept it bare of vegetation.

The impact turned an area of sea floor the size of Belgium into sulfuric acid particles.

There are no known space objects that merit a **Torino score** higher than 1; the chances of an object that rates a 10 coming along during the next century are less than 1 in 1,000.

Events having no likely consequences	0
Events meriting careful monitoring	1
	2
Events meriting concern	3
	4
	5
Threatening events	6
	7
	8
Certain collisions	9
	10

The best known is the **Chicxulub** impact event, 66 million years ago, which marked the end of the Cretaceous era and is thought to have helped wipe out ⅔ of all species, including the dinosaurs. This impact had the force of 100 million megatonnes—if you exploded one Hiroshima bomb for everyone alive on Earth today, you would still be a billion bombs short.

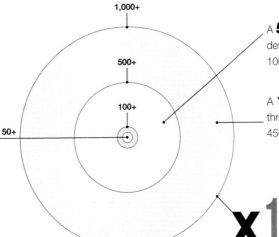

An asteroid with a diameter of **50 m** (165 ft), known as a city-killer because it could level an entire city, hits Earth on average every 750 years—usually in uninhabited regions.

A **500 m**-wide asteroid, big enough to devastate a whole country, hits Earth every 100,000 years, on average.

A **1 km**-wide asteroid, big enough to threaten civilization, can be expected every 450,000 years.

A **10 km**-wide asteroid, known as an extinction-level event, would only be expected every 50–100 million years.

x10

meters diameter

Earth versus the fridge magnets

▶ If Earth were a fridge magnet, you would need 20 of them to stick a piece of paper to your fridge door.

Earth generates a magnetic field, probably because convection and the planet's rotation produce currents in its molten metallic core, which then generate electric currents, which in turn generates a magnetic field. This field is similar to what would be produced by a giant bar magnet running from pole to pole except tilted slightly (by 10° from the axis of spin). The magnitude of the field is weak—about a 20th as strong as a fridge magnet—but it occupies an enormous volume of space, so its total energy is huge.

Magnetic field strength ("magnetic pull") is measured in **tesla** (T) and **gauss** (G)—1 T = 10,000 G. Earth's field strength is roughly 0.5 G, although it varies across the planet—it is strongest at the poles and weakest at the equator.

The strength of a typical horseshoe magnet is ~0.02 T, or 200 G.

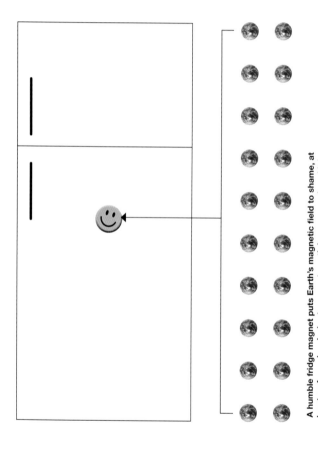

A humble fridge magnet puts Earth's magnetic field to shame, at least as far as local, short-range strength is concerned, and it is considered a more practical alternative to using 20 Earths to stick a shopping list to the fridge door.

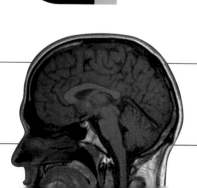

A fridge magnet is around 10 G (0.001 T), while the electromagnet used in an **MRI** (Magnetic Resonance Imaging) machine could be up to 3 T (30,000 G)—3,000 times more powerful than a fridge magnet and 60,000 times more powerful than Earth's magnetic field.

magnetosphere

70,000–300,000km
44,000–186,000miles

In reality, the Earth's magnetic field is deformed into a teardrop shape due to the pressure of the solar wind, a torrent of charged particles from the Sun.

The zone of influence of Earth's magnetic field is called the **magnetosphere**. Typically, the magnetosphere has a radius of about 70,000 km (43,500 miles), but if conditions are right it can balloon out to over 300,000 km (186,400 miles).

Earth's magnetic field has been weakening since it was first measured by Carl Friedrich Gauss in 1836, by around 5% per century, recently accelerating to

5%per decade

Compass needles point to the North Pole (roughly, since the magnetic pole is slightly deflected from the geographic pole). Since the northern tip of the compass needle is a magnetic north pole, the magnetic pole at the North Pole is actually a magnetic south pole.

The U.S. National High Magnetic Field Lab (NHMFL) has a "multi-shot" magnet rated at **90 T**: 90,000 times more powerful than the average refrigerator magnet. The pressure generated inside this magnet is equivalent to 200 sticks of dynamite going off together, or about 30 times the pressure at the bottom of the ocean.

The highest magnetic field generated on Earth is a single-use, destructive pulse magnet at the NHMFL Pulsed Field facility in Los Alamos, which can reach **1,000 T**.

Magnetars—magnetic neutron stars— can have magnetic field strengths of **100billiontesla**

The strength of a magnetic field drops off as the inverse cube of the distance from the magnet, so moving twice as far away causes the field strength to diminish by a factor of **eight**. This means that the field strength of a fridge magnet is effectively undetectable beyond a few millimeters, whereas Earth's field is still powerful thousands of kilometers out into space.

Earth as a Scotch egg

▶ Earth is like a Scotch egg, with a thin crust on top of a thick intermediate layer, and a double-layered core.

A Scotch egg is a savory snack comprising a hard-boiled egg wrapped in sausage meat, all of which is coated in breadcrumbs and deep-fried. A cross-section through a Scotch egg has many similarities with a cross-section through Earth. Both have very thin crusts overlaying intermediate layers that stretch around halfway toward the center—in the Scotch egg it is sausage meat; in Earth it is mantle (partially molten rock). Both have a core composed of two layers—in the Scotch egg, the core is egg white surrounding egg yolk; in Earth, it is an outer core of molten iron and nickel around an inner core of solid iron and nickel.

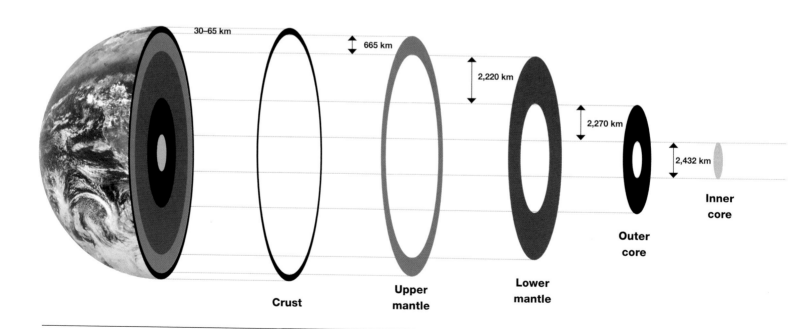

30–65 km

665 km

2,220 km

2,270 km

2,432 km

Crust

Upper mantle

Lower mantle

Outer core

Inner core

Earth's crust consists of ocean floor and continental landmasses, and varies from 10–35 km (6–22 miles) in thickness. The thickest crust, under the Himalaya mountains, is 75 km (46.5 miles). Although this is a lot of rock, compared to the whole planet the crust is extremely thin. If Earth were a basketball, the crust would be about 0.5 millimeters thick.

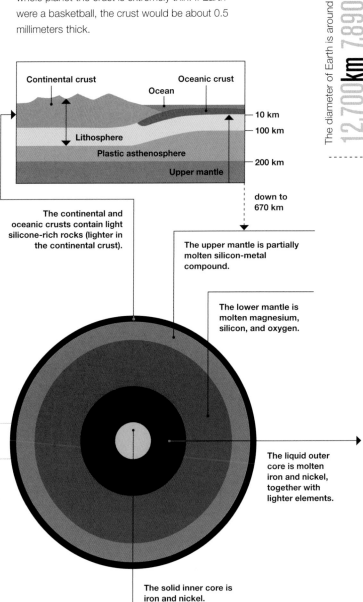

Continental crust

Oceanic crust

Ocean

10 km

100 km

Lithosphere

200 km

Plastic asthenosphere

Upper mantle

down to 670 km

The continental and oceanic crusts contain light silicone-rich rocks (lighter in the continental crust).

The upper mantle is partially molten silicon-metal compound.

The lower mantle is molten magnesium, silicon, and oxygen.

The liquid outer core is molten iron and nickel, together with lighter elements.

The solid inner core is iron and nickel.

The diameter of Earth is around **12,700km 7,890miles**

6,370km 3,960miles

is the distance from the surface of Earth to the center of the core—45 minutes' travel time for a brick falling down a well (except that at the center of Earth the brick would be weightless).

6,000°C

The temperature at the core is thought to be as high as this.

The deepest hole

ever drilled was the Kola borehole in northern Russia. Between 1970 and 1989, Soviet scientists drilled to a depth of 12,262 m (40,230 ft).

 If Earth were an **apple**, the deepest mines would not have broken its skin.

Some of the core is made up of material from a protoplanet that collided with the young Earth 4.5 billion years ago, knocking off material that later coalesced into the Moon.

Before the Moon started slowing down the rotation of Earth through tidal interactions, it used to spin much faster. Even 900 million years ago, Earth's "year" was comprised of 481 "days," with

18 hour-long days

The greenhouse effect

▶ So-called "greenhouse gases" in Earth's atmosphere act like the panes of glass in a greenhouse, letting in light but trapping heat.

Life on Earth depends on the greenhouse effect, a phenomenon first described in the 1820s where gases in the atmosphere act like panes of glass in a greenhouse. Greenhouse glass lets through sunlight, which is absorbed by the air and plants inside the greenhouse, warming them. They then reradiate some of this heat energy as long-wavelength infrared, which cannot pass through the glass but instead is reflected back into the greenhouse. In this way the temperature inside the greenhouse is kept higher than outside.

Similarly, greenhouse gases, such as carbon dioxide, water vapor, and methane, are transparent to visible light but absorb infrared. Instead of passing through the atmosphere and radiating out into space, the heat is locked in, keeping the atmosphere warmer than it would otherwise be. This process has been going on for much of the history of planet Earth, but has become a cause for concern because the overwhelming weight of evidence shows that anthropogenic (manmade) greenhouse gas emissions have caused the carbon dioxide content of the atmosphere to increase sharply, leading to global warming and interfering with the delicate systems of global climate.

Without the greenhouse effect, Earth's average global temperature would be –18°C (–0.4°F), rather than the present 15°C (59°F), and life on the surface of Earth would be impossible.

Anthropogenic carbon emissions currently amount to about 40 billion tonnes/year, just a fifth of the annual natural emissions from sources such as volcanoes and decaying biomass (200 billion tonnes/year). Yet this is enough to tip the delicate balance of the carbon cycle.

20%

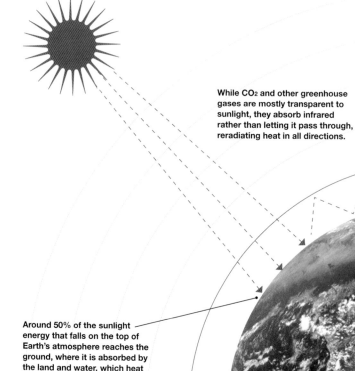

While CO_2 and other greenhouse gases are mostly transparent to sunlight, they absorb infrared rather than letting it pass through, reradiating heat in all directions.

Around 50% of the sunlight energy that falls on the top of Earth's atmosphere reaches the ground, where it is absorbed by the land and water, which heat up and radiate long-wave infrared radiation.

6°C

At current rates of carbon emission, the world could warm by up to 6°C (43°F) over the next few centuries. According to the British Geological Society, it will take Earth 100,000 years to recover.

If atmospheric CO_2 levels can be stabilized at around twice preindustrial levels, the Intergovernmental Panel on Climate Change (IPCC) predicts warming of between

1.4 and 5.8 °C by 2100

The upper limit of this range is more than the difference between the present and the last **Ice Age**, while even the lower end of the range would be the biggest temperature change in the entire history of civilization.

One consequence of global warming could be increased melting of the ice caps leading to rising sea levels. If all the ice sheets melted, **sea levels** would rise by 60 m (200 ft), roughly the height of a 20-story building.

As a conservative prediction, the IPCC estimates that sea levels will rise by between

9 and 88 cm by 2100

If the upper estimate proves to be correct, at least **a fifth** of Bangladesh will be under water.

Around 75% of the world's population lives within 80 km (50 miles) of the sea.

Aurora Borealis, the "Northern lights," appear in these two outermost areas of the atmosphere.

Exosphere

Thermosphere

Mesosphere

Stratosphere

Troposphere

The exosphere, which extends beyond 800 km (500 miles) above Earth's surface, is more outer space than atmosphere. Gas molecules may be hundreds of miles apart at this level.

The thermosphere extends from around 85 km (50 miles) above Earth's surface to around 800 km (500 miles), depending on solar activity. This is the layer in which the International Space Station orbits; above 100 km (60 miles) is widely considered to be space. Temperatures can be extremely high, but the air is so rarefied that the normal definition of temperature no longer applies.

The mesosphere extends from the top of the stratosphere to around 85 km (50 miles) above Earth. This is the layer where meteorites burn up. It is extremely cold.

The stratosphere, extending from the top of the troposphere to around 50 km (30 miles) above Earth, includes the ozone layer.

The troposphere contains the vast majority of gas in the atmosphere, and extends from the surface of Earth to a height of 10 km (6 miles) at the poles and 20 km (12 miles) at the equator. Airplanes and weather are confined to the troposphere.

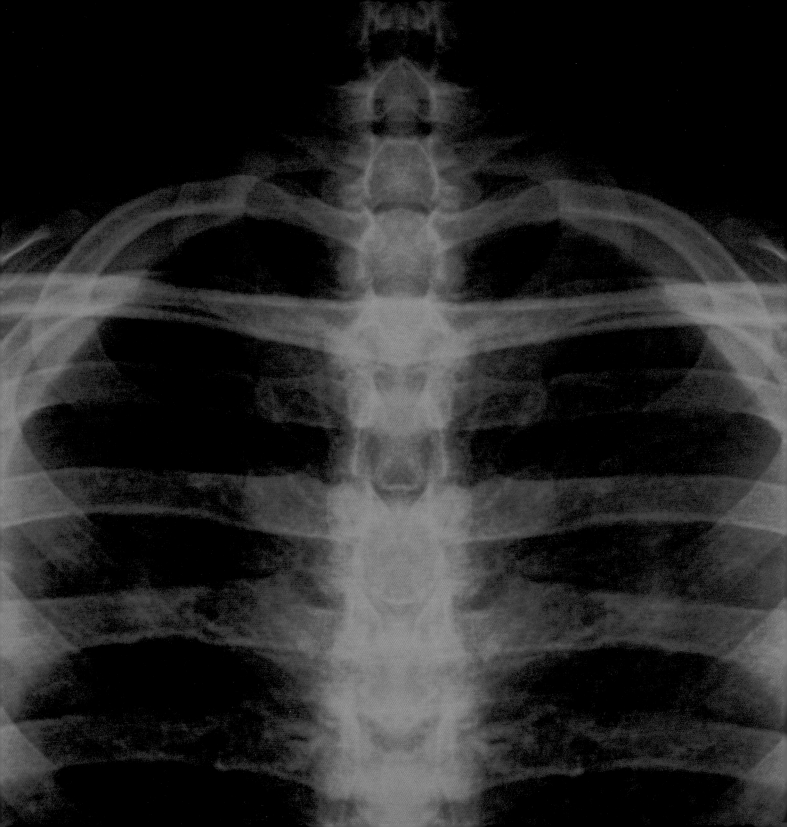

Section Six

▶ The human body seems like familiar territory, yet it can
be hard to grasp the incredible capabilities of our organs
and tissues, and the amazing facts and figures describing
its complex makeup, without the help of analogies.
Thought experiments are also important in exploring
philosophical issues concerning the body and brain.

The Human
Body

What are you made of?

▶ If your body were broken down into its elemental components, there would be a cube of oxygen the size of a small television, a brick-sized chunk of charcoal, a kilogram of calcium, and a teaspoon of iron.

The majority of naturally occurring elements are found in the human body, though mostly in trace amounts only. Given that life is based on organic chemistry and that the body is mainly water, it is to be expected that the most frequently occurring elements in the body are oxygen, hydrogen, and carbon. But dozens of obscure, toxic, and even radioactive elements are also present, many of them with unknown (or without any) biological functions.

Data from Ed Uthman MD gives the mass of each element present in a 70 kg (154 lb) individual, together with the size of the cube it would make if gathered together.

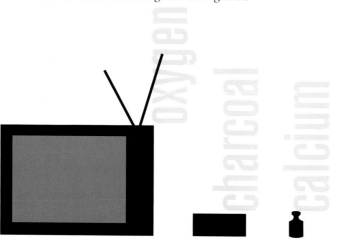

oxygen charcoal calcium iron

The human body is about 60% water, although the exact amount depends on age, gender, and fat content. Babies are about 78% water at birth, while a one-year-old child is around 65% water. Adult men are typically 60% water, but since fat contains less water than muscle, and women have a higher proportion of fat in their bodies, adult women are typically around 55% water.

This chart shows the mass of each element present in the body, and the size of cube it would make if extracted and purified into its pure elemental form. Bear in mind that gaseous elements have much lower density and therefore take up large volumes in relation to their mass.

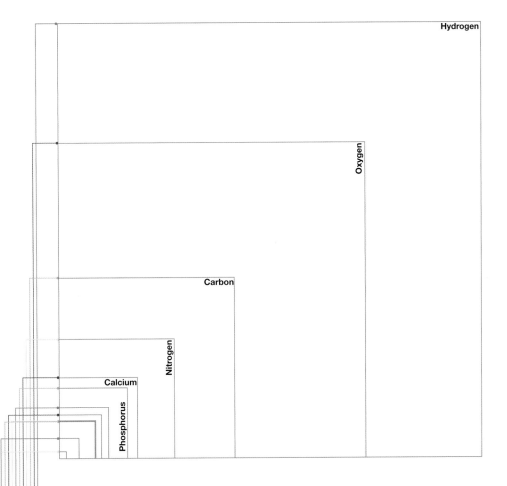

Rubidium is the most abundant element in the body that has no known role, despite the fact that an average person has 0.68 g of it, making it the 16th most abundant element in the body.

At 0.11 mg, **vanadium** is the least abundant element in the body that has a known role. None of the elements below vanadium in the table have known biological functions.

The body isn't actually made up of elemental atoms; it is almost entirely composed of molecules. For instance, there are over 200,000 types of protein in the human body, fewer than 2% of which are comprehensively **understood**

2%

ELEMENT	MASS	CUBE SIZE	ELEMENT	MASS	CUBE SIZE	ELEMENT	MASS	CUBE SIZE
oxygen	43 kg	33.5 cm	aluminum	60 mg	2.8 mm	silver	2 mg	0.6 mm
carbon	16 kg	19.2 cm	cadmium	50 mg	1.8 mm	niobium	1.5 mg	0.6 mm
hydrogen	7 kg	46.2 cm	cerium	40 mg	1.7 mm	zirconium	1 mg	0.54 mm
nitrogen	1.8 kg	12.7 cm	barium	22 mg	1.8 mm	lanthanium	0.8 mg	0.51 mm
calcium	1.0 kg	8.64 cm	iodine	20 mg	1.6 mm	gallium	0.7 mg	0.49 mm
phosphorus	780 g	7.54 cm	tin	20 mg	1.5 mm	tellurium	0.7 mg	0.48 mm
potassium	140 g	5.46 cm	titanium	20 mg	1.6 mm	yttrium	0.6 mg	0.51 mm
sulfur	140 g	4.07 cm	boron	18 mg	2.0 mm	bismuth	0.5 mg	0.37 mm
sodium	100 g	4.69 cm	nickel	15 mg	1.2 mm	thallium	0.5 mg	0.35 mm
chlorine	95 g	3.98 cm	selenium	15 mg	1.5 mm	indium	0.4 mg	0.38 mm
magnesium	19 g	2.22 cm	chromium	14 mg	1.3 mm	gold	0.2 mg	0.22 mm
iron	4.2 g	8.1 mm	manganese	12 mg	1.2 mm	scandium	0.2 mg	0.41 mm
fluorine	2.6 g	1.20 cm	arsenic	7 mg	1.1 mm	tantalum	0.2 mg	0.23 mm
zinc	2.3 g	6.9 mm	lithium	7 mg	2.4 mm	vanadium	0.11 mg	0.26 mm
silicon	1.0 g	7.5 mm	cesium	6 mg	1.5 mm	thorium	0.1 mg	0.20 mm
rubidium	0.68 g	7.6 mm	mercury	6 mg	0.8 mm	uranium	0.1 mg	0.17 mm
strontium	0.32 g	5.0 mm	germanium	5 mg	1.0 mm	samarium	50 µg	0.19 mm
bromine	0.26 g	4.0 mm	molybdenum	5 mg	0.8 mm	beryllium	36 µg	0.27 mm
lead	0.12 g	2.2 mm	cobalt	3 mg	0.7 mm	tungsten	20 µg	0.10 mm
copper	72 mg	2.0 mm	antimony	2 mg	0.7 mm			

Emptying a bathtub with a teacup

▶ To see what it's like to be your own heart, try using a teacup to empty a bathtub in 15 minutes—then do it again and again, without stopping, for the rest of your life.

Your heart and circulatory system are remarkably engineered to distribute effectively the amazing soup of cells collectively termed blood. The facts and figures relating to this system are breathtaking, with trillions of cells hurtling through thousands of kilometers of blood vessels thousands of times a day.

To get a clearer picture of the role of the heart and circulatory system, imagine the body as a car. The heart is the engine, generating the power to drive the body, while the circulatory system is the driveshaft, distributing power to the moving parts. In fact it distributes power in the form of oxygenated blood to every one of the c.100 trillion cells in your body, with the exception of the corneas, the only tissues without a blood supply.

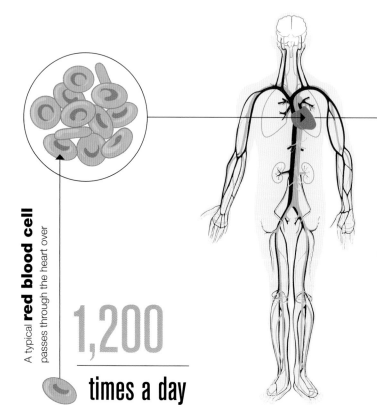

Every day the heart **expends enough energy** to drive a truck 32 km (20 miles)—over a lifetime it could power a truck to the Moon and back.

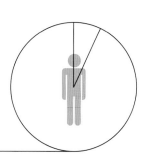

The smallest vessels in the circulatory system are **capillaries**—some are only as wide as a single red blood cell. Capillaries in the human body number around

10 billion

A typical **red blood cell** passes through the heart over

1,200 times a day

7% of your body weight is accounted for by blood.

1 milliliter

of human blood contains about

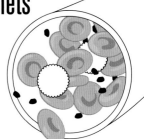

5,000,000 red blood cells

5-10,000 white blood cells

2-300,000 platelets

You have over **50 billion** white blood cells in your bloodstream.

Platelets are cell fragments that help with clotting.

The human circulatory system is a vastly complex network of vessels that carries oxygenated blood from the lungs to the body's extremities.

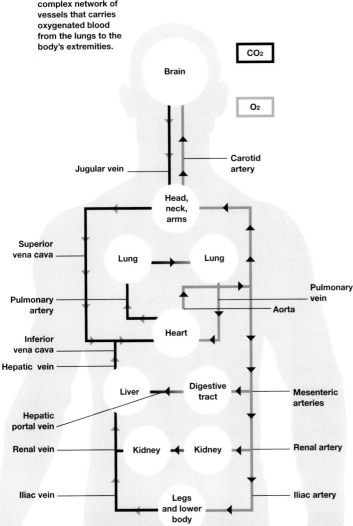

 CO₂

 O₂

Brain

Jugular vein — — Carotid artery

Head, neck, arms

Superior vena cava

Lung — Lung

Pulmonary artery

Pulmonary vein

Aorta

Inferior vena cava

Hepatic vein

Heart

Liver — Digestive tract

Hepatic portal vein

Mesenteric arteries

Renal vein

Kidney ← Kidney

Renal artery

Iliac vein

Legs and lower body

Iliac artery

10 billion

blood cells are produced by a child every day.

In an adult, blood cells are produced in the **bone marrow**. You have about 1–2 liters (1¾–3½ pints) of bone marrow, which produces red blood cells at a rate of 3 million/sec, although this can increase to a startling 30 million/second when needed.

Red blood cells account for a third of all the cells in your body. They are donut-shaped to give them a higher surface area. It is estimated that the total surface area of all the red blood cells in an average adult is 3,800 m² (4,500 square yards)—enough to cover 14 tennis courts.

White blood cells are part of your immune system; they are specialized to fight off invading bacteria, viruses, and other threats, but they need training. The thymus is a gland situated between the lungs, which acts as a boot camp for these cells. Only 5% of cells that enter the thymus for training make it out the other end.

As shown in the previous entry, the human circulatory system is a truly amazing thing, although we rarely notice it until it fails. We certainly do not often recognize just how remarkable are the feats of strength that keep blood circulating through our bodies at all times. The heart is always pumping; whether we are running a marathon or sleeping motionless, blood is flowing through our veins, continually pumped by that greatest of all the body's muscles. The heart never rests, never sleeps, never stops using energy, and over the course of a lifetime it accomplishes some fairly spectacular feats, pumping billions of times, pushing millions of gallons of blood through the body, all while making use of tremendous amounts of energy.

The circulatory system is diverse; the aorta, the largest artery in the body, is almost the diameter of a garden hose. Capillaries, on the other hand, are so small that it takes ten of them to equal the thickness of a human hair.

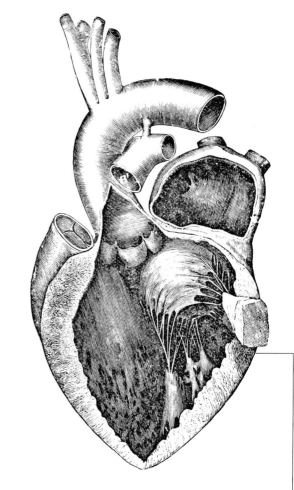

Swimming pools and supertankers

▶ It would take your heart about one year to fill an Olympic-sized swimming pool, and during an average lifetime it pumps about 1.25 million barrels of blood— enough to fill three supertankers.

The heart is not really that large. In fact, a child's heart is about the same size as your fist. For an adult, it's about the same size as two fists.

x2¹/₂

If all the capillaries in a human body were laid out end to end, they would wrap around the world two and a half times.

6 liters

is the amount of blood in an average human adult.

Blood circulates through the body about once every 90 seconds. In one day, the blood travels a total of 4,800 km (3,000 miles). That's the distance across the USA from coast to coast.

The heart beats about 100,000 times each day (that's a little more than once per second) and about 35 million times in a year. During an average lifetime, the human heart will beat more than 2.5 billion times.

100thousand heartbeats per **day**

35million heartbeats per **year**

2,500million heartbeats per **lifetime**

The heart uses roughly the same amount of **force** to pump blood out to the body as the human hand giving a strong squeeze. Even at rest, the muscles of the heart work twice as hard as the leg muscles of a person sprinting.

Bones of steel

▶ If your bones were made of metal, you could be four times taller, 64 times heavier, and able to eat rocks.

Biological materials in general, and human bone in particular, have been engineered through evolution over millions of years. This gives them a mixture of properties, so that in some respects they outperform artificial materials, and in others they do not quite match up.

Bone is a living tissue composed of long protein collagen fibers interwoven with crystals of calcium salts (with similar properties to marble), suffused with living cells and blood vessels. The collagen fibers have great tensile strength (they can endure stretching forces), while the calcium salt crystals have compressional strength (they can endure crushing), so that bone as a composite material combines both properties. In fact, bone is structured along similar principles to steel-reinforced concrete, with the collagen as the steel and the calcium salts as the concrete.

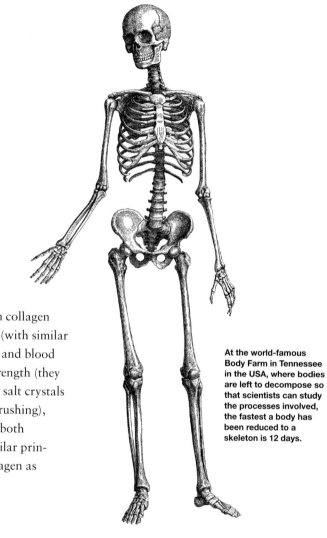

At the world-famous Body Farm in Tennessee in the USA, where bodies are left to decompose so that scientists can study the processes involved, the fastest a body has been reduced to a skeleton is 12 days.

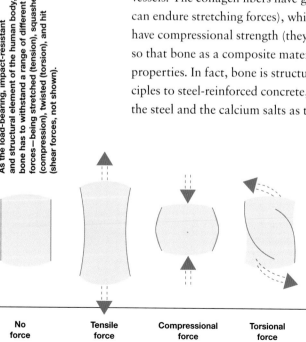

As the load-bearing, impact-resistant and structural element of the human body, bone has to withstand a range of different forces—being stretched (tension), squashed (compression), twisted (torsion), and hit (shear forces, not shown).

No force	Tensile force	Compressional force	Torsional force

Compared to **reinforced concrete**, bone has greater compressional strength and almost as good tensile strength.

Compared to **raw concrete**, sources differ, rating human bone as 4–40 times stronger than concrete.

1/3 In terms of tensile strength, bone is similar to **cast iron** but only weighs a third as much.

Bone has to be strong in order to endure the **stresses** placed on it by muscle contraction and the body's weight.

A **running man** exerts force on his bones equivalent to a dead weight of **270kg**

Even bone is no match, however, for the best artificial materials. Some **steel alloys** have ten times the tensile strength and fracture toughness (the ability to resist fracturing when there is a crack).

x10 better

x25 more energy Also, steel can absorb 25 times more energy than bone without fracturing.

A bone made from **titanium alloy** would be only about 1.3 times heavier than real bone, but would be five times stronger. Such bones would never need repairing, because the alloy's fatigue strength (the strain it can bear repeated over long periods) is five times higher than anything a real bone would experience. In an accident, a titanium alloy would simply bend and could be bent back into shape afterwards.

A baby has 270 bones, while an adult has only 206 bones. The baby's soft bones **fuse** to give the final number.

270

206

x0.25 **Chitin**, the carbohydrate used by insects for their hard bodies, gives them extraordinary properties. The largest land animal, the African elephant, can carry only around a quarter of its own weight. The horned dung beetle, *Onthophagus taurus*, can pull 1,141 times its own body weight—equivalent to a person lifting six full double-decker buses. **x1,141**

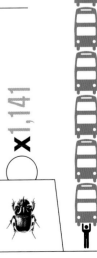

Outnumbered

▶ The bacteria that live in and on your body outnumber your own cells by more than ten to one.

Your immune system works overtime to ensure that bacteria are kept out of your internal tissues, but your external surfaces—which include your mouth, throat, stomach, and intestines—are in constant contact with the environment, a world teeming with microbes. It is not feasible to keep these bugs away, so instead the human body has evolved to make the best of the situation, by living in a mutual symbiotic relationship with them.

This "mutualism" describes the symbiotic relationship between you and your normal flora (as your microbial population is known). Your flora gorge themselves on the food you eat, the substances you secrete and the cells you shed, and in return many of them provide essential nutrients, such as vitamins the body cannot make for itself. Your gut flora in particular make up a vital part of your digestive system, the collection of organs that turns food into nutrients that your cells can handle.

Every time you sit down to eat, you literally share your dinner table with a range of bacteria. All household surfaces host some microorganisms, the vast majority of which are harmless.

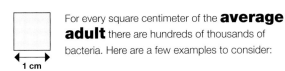

For every square centimeter of the **average adult** there are hundreds of thousands of bacteria. Here are a few examples to consider:

1 billion in your mouth

10 million in your armpit

1 trillion on your entire skin

100,000 per cm²

10 million in your groin

You have at least 750 trillion gut bacteria, belonging to more than **400** different species.

750 trillion

In fact there may be far more species than previously believed. A recent experiment looking at the RNA "fingerprints" of bacteria in the mouth suggested that a mere 1% of the species present had ever been **identified by science**.

1%

The human gut is 8 m (26 ft) long and contains bacteria that break down molecules that the human body cannot process itself. It is the work of these bacteria that produces gases inside the intestines.

You produce 1–1.5 liters (1.75–2.6 pints) of **saliva** a day.

There are **10,000** taste buds on the tongue.

Stomach acid can burn skin and dissolve razor blades—it is 1,600 times more acidic than vinegar. You produce about 1.5 liters (2.6 pints) a day.

You can lose up to 80% of your liver and regenerate it within **10 days**.

Your liver cleans 1 liter (1.75 pints) of blood every minute—during a year it processes enough blood to fill 23 milk tankers.

Dead head

▶ If you have shoulder-length hair, a piece of clothing the same age would be threadbare and ragged; if you bleach or dye your hair regularly, a piece of clothing the same age would have dissolved months ago.

Every hair on your head is dead tissue composed of much the same stuff as a corn on your foot or callus on your hand. Animal horns too are often made of the same material—keratin. A hair is a shaft of keratin fibers coated with shinglelike, keratin-filled dead cells. Hair grows pretty slowly, so by the time a strand of hair reaches shoulder length it is usually around two years old. Few pieces of fabric would survive this length of time if exposed to the same treatment as a head of hair: repeated washing (perhaps daily), blow drying, styling with hot irons, spraying and stiffening with products, and possibly bleaching and dyeing on a regular basis, all in addition to the natural wear and tear produced by exposure to the wind, rain, cold, and sunlight, and the physical battering caused by brushing, curling, pinning, and simply rubbing against things.

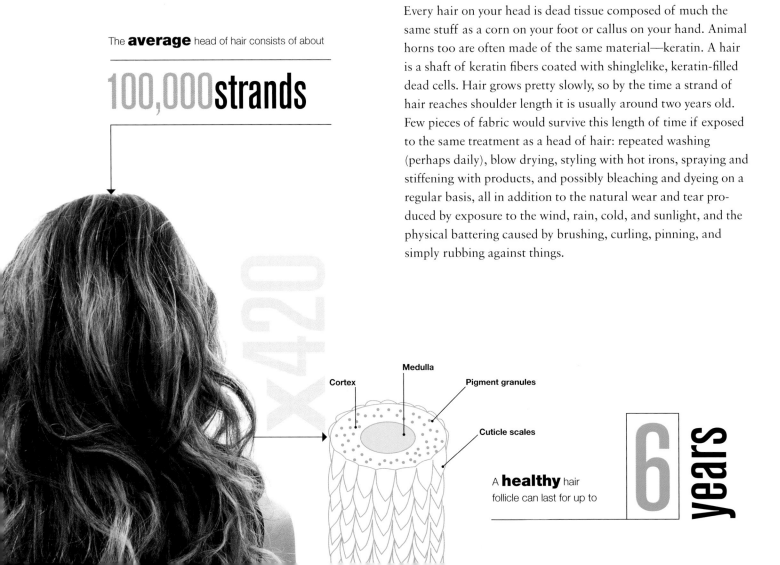

The **average** head of hair consists of about

100,000 strands

x420

Cortex · Medulla · Pigment granules · Cuticle scales

A **healthy** hair follicle can last for up to

6 years

According to Clarence R. Robbins, author of the 2002 book *Chemical and Physical Behavior of Human Hair*:

16

The rate, in centimeters, that scalp hair grows per year. This is around 6 inches a year.

14

Hair around the temples grows more slowly—around 14 cm (5.5 in) a year.

10

Beard hair grows at just 10 cm (3.9 in) a year.

Hair growth slows with **age**, and can reduce to just 3 cm (1.2 in) a year.

3

centimeters per year

hair
vs
nails

A fingernail will grow by about

3

centimeters per year

In a race between hair and fingernails, hair is definitely the hare. **Fingernails** grow at the rate of about one nanometer per second—roughly the width of an average molecule.

1

nanometer per second

Fingernails grow at about

0.09

roughly the diameter of a human hair.

millimeters per day

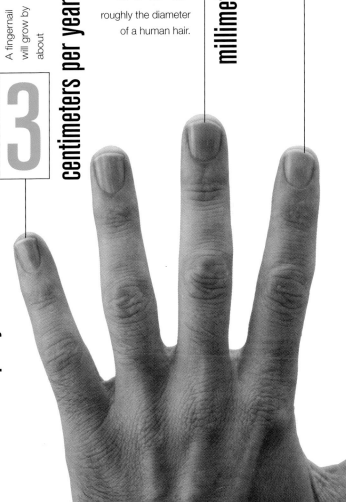

Swimming in molasses

▶ A human spermatozoa swimming toward its target is like a man in a swimming pool full of molasses, unable to move any part of his body faster than one centimeter per minute.

Human sperm have a tough job. They are blasted into a hostile, acidic environment, and then have to swim 20,000 times their own body length through thick mucus to reach a target 85,000 times bigger than they are, only to discover that hundreds of other sperm have got there first. What makes it particularly tough for sperm is that at such microscopic scales the viscosity of the medium through which it is swimming has a massive effect on it, as expressed through something called its Reynolds number. The lower the number, the greater the challenge. A man swimming in water has a Reynolds number of around 10,000; for a sperm it's nearer to 0.00003. If a man had to swim at the same Reynolds number, it would be like swimming in molasses while only able to move his limbs incredibly slowly.

The average healthy ejaculation contains up to

150 million sperm

enough to populate Bangladesh or Nigeria, if all the sperm were viable.

A healthy male ejaculation has a sperm count of at least 20 million sperm per milliliter; if conditions are right, some of the tens of millions of sperm ejaculated may reach their target, but only one can penetrate it—as soon as the first sperm breaches the membrane of the egg, biochemical changes take place to block any others.

A sperm has to cover about **10–18 cm** (4–7 in) between its "jump off" point and the egg.

The tail of a sperm is 55 microns long, but its body is just 5 microns long (0.005 mm), making the journey

20,000–36,000 body lengths

0 1 2 3 4 5 6 7 8 9 10 11 12 13 14 15 16

If the sperm were the size of a **salmon**, its journey would be more than

14 km 8.6 miles

If it were the size of a **sperm whale** its journey would be over 380 km (236 miles).

380 km 236 miles

10 blps

A sperm travels at about 1–4 millimeters a minute—in body lengths per second (see pp. 124–5), this is 10 blps. If the sperm were a human it would be moving at 18 mps (60 ft/s), or 65 km/h (40 mph)—particularly impressive given its low Reynolds number.

65 km/h
40 mph

Semen is packed full of **nutrients**, including vitamin C, calcium, chlorine, cholesterol, citric acid, creatine, fructose, lactic acid, magnesium, nitrogen, phosphorus, potassium, sodium, vitamin B12, and zinc.

The caloric content of the average ejaculation is 5 to 25 calories; its **protein** content is roughly equivalent to the egg white of a large egg.

5–25 calories

Some studies found that between 15 and 20% of young men have **sperm counts** of less than 20 million per milliliter. A dairy bull has a sperm count in the billions.

A widely cited paper by Carlson *et al.* in the *British Medical Journal* in 1992 suggested that human sperm counts had **dropped 40%** over the previous 50 years, although this and similar findings are unproven.

The Ship of Theseus

▶ You are like the Ship of Theseus—every single molecule of your body has been replaced, so are you still you?

The Ship of Theseus is a philosophical paradox based on a relic supposedly preserved by the ancient Athenians. According to the 1st-century Greek philosopher Plutarch:

"The ship wherein Theseus and the youth of Athens returned had thirty oars, and was preserved by the Athenians down even to the time of Demetrius Phalereus [c.300 BCE], for they took away the old planks as they decayed, putting in new and stronger timber in their place, insomuch that this ship became a standing example among the philosophers, for the logical question of things that grow; one side holding that the ship remained the same, and the other contending that it was not the same."

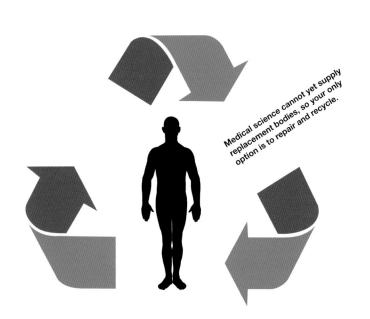

Medical science cannot yet supply replacement bodies, so your only option is to repair and recycle.

In other words, if you replace every part of something, is it the same thing that it was before or something new? This paradox concerning continuity of identity could be said to apply to humans as well. Most cells in the human body are replaced at least once a month, and even those that are not are constantly having their constituent parts— proteins and carbohydrates, for example—recycled, repaired, and replaced. Given that there is barely a single molecule in your body that was part of you 15–20 years ago, are you still the same person you were then? It may seem obvious that the answer is yes, but what if someone took all the discarded cells and molecules and reconstituted an exact copy of you with them? Which "you" would be the more authentic one?

Most living cells in the human body are less than a month old; exceptions include liver cells (which last for years) and brain cells (which last for a lifetime).

The components of each brain or liver cell are constantly **renewed**, however, so that no individual component is likely to be more than a month old.

You are born with more brain cells than you will ever have again—around

100,000,000,000

According to one estimate, you lose

500

brain cells
per hour

9
years old

Almost **no** molecule in an adult human body is likely to be more than

The cells in the top layer of your **skin** are all dead—you are entirely covered in dead cells. An average adult is wearing over 2 kg (4.4 lb) of dead skin cells, billions of which fall off each day.

2 kg

ATP

DNA replication is phenomenally **accurate**, even when compared to the best human technology. Proofreading enzymes help to keep error rates as low as one base in every billion—sometimes even one base in every ten billion. This is like typing out the complete works of Shakespeare 1,400 times over and making just one mistake.

The most spectacular example of the high turnover rate of molecules in the human body is the quantity of ATP you get through. ATP is the molecule the cell uses for energy—the physiological equivalent of the electron in an electrical circuit.

Each cell in your body has around

a billion ATP molecules,

which are used up and replaced every two minutes.

Thanks in part to this accuracy, **malignancies** occur just once every 100,000 trillion cell divisions in humans.

1 in

100,000 trillion

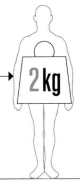

Every day you produce and use up a volume of ATP equivalent to

half your body weight

Descartes' evil genius

▶ If you were nothing more than a brain in a vat of fluid, wired up to a computer that simulated reality for you, would you be able to tell?

In 1641 French philosopher René Descartes (1596–1650) posed an argument that challenges the very nature of reality, posing the ultimate doubt about the universe. "Suppose," he wrote, "some evil genius not less powerful than deceitful, has employed his whole energies in deceiving me; I shall consider that ... all ... external things are but illusions and dreams of which this genius has availed himself to lay traps for my credulity." In other words, if the only means we have to determine the nature and existence of reality is the information that arrives in the mind via the senses, it is possible that we could be deceived via these senses.

The modern version of this skeptical dilemma is the "brain in a vat" problem. If you are not really a person with a body, walking around and interacting with reality, but actually just a brain suspended in a vat, fed a simulation of reality by a powerful computer, there would be no way to tell the difference. This conceit forms the basis of the film *The Matrix*.

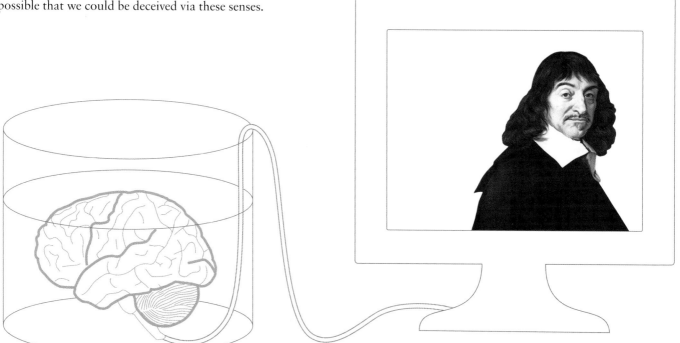

The **cerebrum**—the outer part of the brain—is thought to be the seat of the conscious mind. If the cerebrum's wrinkles were all unfolded, it would cover four pieces of A4 paper.

90 minutes

Adults generally cycle through the five phases of sleep every 90 minutes or so (known as an ultradian rhythm), although the time spent in each phase changes over the course of the night, and over the course of the lifespan.

6 minutes.

The mind can have extraordinary powers over the body. Hindu holy men, known as **saddhus**, can consciously slow their heart rates to just two beats a minute and stay under water for up to six

Some Tibetan monks practice a skill known as **tumo**, where they learn to raise the temperature of their fingers and toes by up to 8°C (15°F), simply by an effort of will.

When you first drop off, you spend more time in deep sleep and only a couple of minutes in **REM** sleep, but by the next morning your REM phases may be up to as long as

30 minutes

An everyday version of the evil genius scenario happens during **dreaming**, when your subconscious feeds sensory simulations, aka dreams, to your sleeping mind.

You spend the equivalent of

122 days

asleep every year—that's about a **third** of your life.

X 1,825 dreams

Most people dream around five times during each eight-hour period of sleep. Based on this number, you have about 1,825 dreams **every year**.

The dolphin and the echidna are the only mammals that do not have REM sleep (as far as researchers can tell).

Newborn babies spend up to **70%** of their sleep time in REM sleep.

Shark bites giant

▶ If a globe-spanning giant with his head in Baltimore and his toe off the South African coast were bitten on the foot by a shark on Monday, he would not feel the bite until Wednesday and would not react until Sunday.

This is an analogy used by Johns Hopkins University neuroscientist David Linden to illustrate the relative slowness of human nerves. In the nervous system, nerve cells, known as neurons, transmit messages via electrical impulses that travel along projections known as axons. When you are bitten on the foot, an impulse travels to your brain from the pain receptors in your foot, via neurons with very long axons that connect your foot to your brain via the spinal cord. When the impulse arrives in the brain it triggers the conscious perception of pain, but it also triggers an instinctive muscular reaction through a relatively simple processing loop in the spinal cord. The speed of nerve impulses and mental processing seems fast, but if a human were scaled up to gigantic size the relative slowness of the process would be exposed.

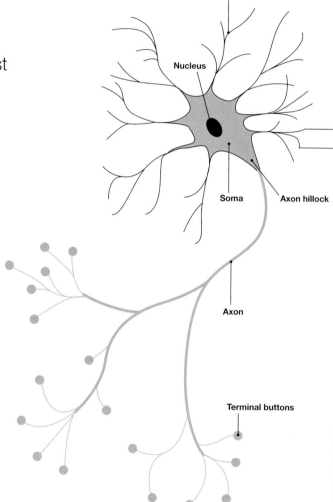

An electron micrograph of the cell body—or soma—of a neuron, showing the axon and a number of dendrites branching off it.

One neuron may be linked to as many as **50,000** others; the brain is estimated to contain over 100 trillion connections. The total number of possible connections is higher than the number of atoms in the universe.

100 trillion+

Axons can stretch up to **1 m** (3.3 ft) to make connections with other neurons. If the cell body of a single neuron were enlarged to the size of a tennis ball, the axon would be well over half a mile long.

Impulses can travel along nerve fibers at speeds of up to

270milesperhour

20%

Oxygen

Nerve cells demand so much energy that the brain uses up to 20% of the body's oxygen.

Glucose

The brain consumes an incredible 60% of the body's glucose supply.

60%

The brain is like an aristocratic household—the pampered upper-class neurons have a staff of helper glial cells to tend to them.

It is commonly reported that glial cells outnumber neurons by at least ten to one. But recent research by Suzana Herculano-Houzel of the Universidade Federal do Rio de Janeiro and others shows that this is a myth, and that the ratio is nearer one to one.

The average adult male human brain, weighing **1.5 kg** (3 lb), contains

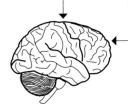

86 billion neurons

85 billion glial cells

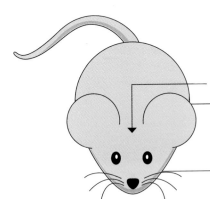

Rodent brains are built differently from human ones.

 A rodent with the same number of neurons as a human would weigh 50 tonnes and have a 35 kg (77 lb) brain.

If humans had a brain built like a rat's, it would weigh only **145 g** (5 oz) and be made up of only

12 billion neurons

In camera

▶ Both the eye and a camera have adaptive lenses that focus an image via an aperture onto a light-sensitive surface.

An "old-fashioned," non-digital camera uses a lens to gather and focus light, an aperture to control the amount of light admitted and thus control for different brightness conditions, a black-lined interior to make sure there are no internal reflections, film to capture an inverted image at the back of the camera, and a lens cap to protect the lens when not in use. The eye uses two lenses to gather and focus light; one of them, the cornea, is fixed, but the other can adapt (by changing shape) to focus on near and far objects, just as a camera lens can focus on near and far objects (by moving backward or forward). The eye has an aperture to control brightness (the iris), and is black inside to prevent reflections. It has an eyelid to protect the lens when not in use, and an inverted image is captured by the retina at the back of the eye.

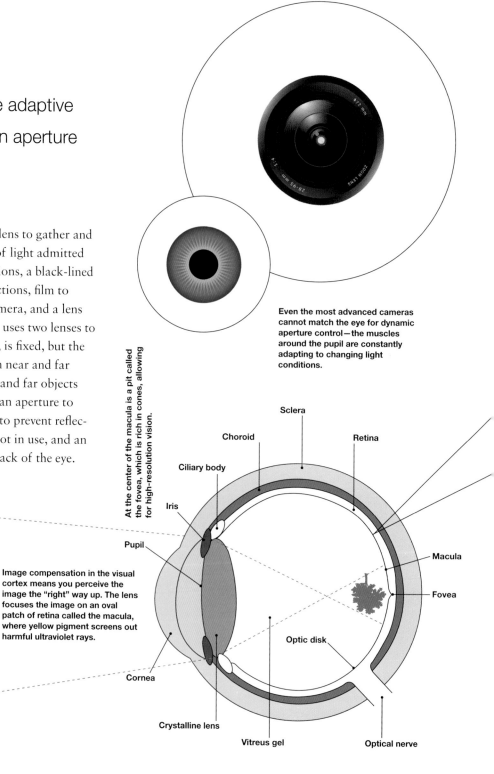

Even the most advanced cameras cannot match the eye for dynamic aperture control—the muscles around the pupil are constantly adapting to changing light conditions.

At the center of the macula is a pit called the fovea, which is rich in cones, allowing for high-resolution vision.

Image compensation in the visual cortex means you perceive the image the "right" way up. The lens focuses the image on an oval patch of retina called the macula, where yellow pigment screens out harmful ultraviolet rays.

Light reflects off an object, is gathered by the cornea and focused by the lens, and falls on the retina as an inverted image.

Sclera

Choroid

Retina

Ciliary body

Iris

Pupil

Macula

Fovea

Optic disk

Cornea

Crystalline lens

Vitreus gel

Optical nerve

The retina contains two types of
photoreceptors

Counterintuitively, light must pass through nerve cells and other layers to reach the rods and cones that actually detect it. Possibly this is because the rods and cones need to be near the blood supply at the back of the retina.

Rods for black-and-white vision

Cones for color vision

There are 120 million rods and 6–7 million cones. The cones are concentrated in the center of the retina.

The rods are over a thousand times more sensitive than the cones. According to some sources, under the right conditions they are so sensitive that a single photon can trigger one.

Cone   Rod

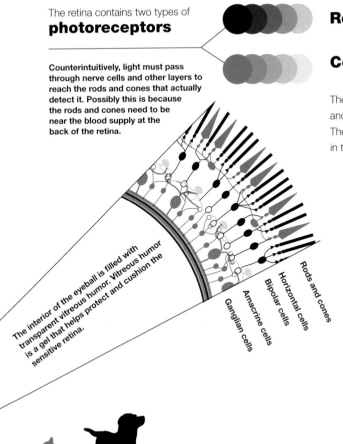

The interior of the eyeball is filled with transparent vitreous humor. Vitreous humor is a gel that helps protect and cushion the sensitive retina.

Ganglian cells
Amacrine cells
Bipolar cells
Horizontal cells
Rods and cones

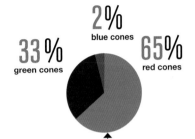

2% blue cones

33% green cones

65% red cones

The cones can be divided by the **color** sensitivity of the pigments they carry.

It is a **myth** that cats and dogs can see only in black and white, although it is true that while humans have three color pigments in their eyes, dogs and cats have only two, so they are like red–green color-blind people.

The mantid shrimp is the world champion of color vision. It has more visual pigments than humans, and also has color filters in its eye. Combining the two means it effectively has 16 color pigments, compared to our three.

color pigments

16

An eagle's retina has about
600,000 cones
per square millimeter, four times as many as the human eye.

1mm²

X 4

An eagle's eyesight is at least four times better than that of a person with perfect vision. An eagle can identify a rabbit moving almost a mile away. An eagle flying at an altitude of 300 m (1,000 ft) over open country can spot prey over an area of almost 7.75 square kilometers (3 square miles).

Section Seven

▶ Science and technology have made possible
achievements of extraordinary grandeur and
mind-boggling scale—sometimes the only
adequate way to appreciate them is through
the lens of analogy, setting them in the
context of more familiar concepts.

Technology

A Manhattan every 18 months

▶ To cope with their astronomical growth, over the next 15 years Chinese cities are planning to construct skyscrapers equivalent to ten Manhattans.

The explosive rate of urbanization in China, combined with planning laws to restrict sprawl and protect farmland, means that an estimated 50,000 new skyscrapers will be built in Chinese cities over the next 15 years, according to the consultant McKinsey & Co. This is equivalent to ten Manhattans.

50+ stories

According to the Council on Tall Buildings and Urban Habitat, 294 buildings of 50 stories or taller were built in the **last decade**—more than all of the skyscrapers constructed in the century before that.

In the next **15 years** China's urban population will expand by the equivalent of the population of the USA.

About **half** of those buildings were in China and Hong Kong.

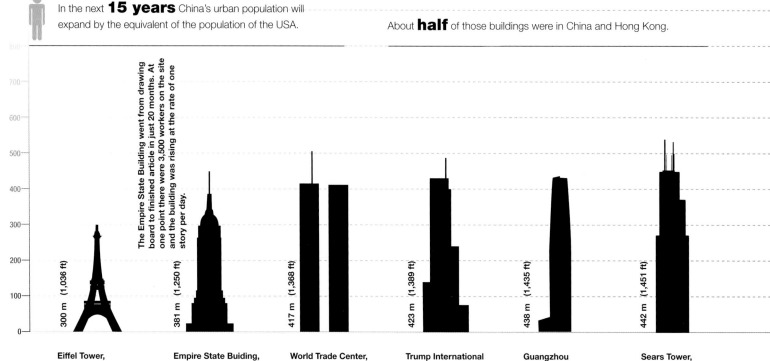

The Empire State Building went from drawing board to finished article in just 20 months. At one point there were 3,500 workers on the site and the building was rising at the rate of one story per day.

Eiffel Tower, France 1889 — 300 m (1,036 ft)

Empire State Buiding, USA 1931 — 381 m (1,250 ft)

World Trade Center, USA 1973-2001 — 417 m (1,368 ft)

Trump International Hotel and Tower, USA 2009 — 423 m (1,389 ft)

Guangzhou International Finance Center, China 2010 — 438 m (1,435 ft)

Sears Tower, USA 1974 — 442 m (1,451 ft)

X5.5

Five and a half times taller

The tallest building—in fact the tallest human structure ever built—is the **Burj Khalifa** in Dubai, which is 828 m (2,716.5 ft) and more than 160 stories high. It is more than twice as high as the Empire State Building.

6 years

It took six years, 39,000 tonnes of reinforcing steel, 330,000 cubic meters of concrete, and 22 million man-hours to complete. The reinforced piles of the foundations alone contain more than 110,000 tonnes of concrete, buried up to 50 meters deep.

The first skyscraper to top 300 m was the **Chrysler Building**, completed in 1930—it was 319 m (1,046 ft), although it was surpassed almost immediately by the Empire State Building, which ruled as the world's tallest building from 1931 to 1971, at 381 m (1,250 ft).

The Burj Khalifa is only half the height of the Illinois, a mile-high skyscraper proposed by Frank Lloyd Wright in 1956. His design included 76 atomic-powered, quintuple-deck lifts, including express elevators that could reach the top of the **528-story** building in a minute. Unfortunately, these would have taken up most of the floorspace.

The Burj Khalifa is five and a half times taller than the Great Pyramid at Giza in Egypt, which was the world's tallest building for over **4,000 years**. Not until around 1311 was it knocked from its lofty perch by the spire of Lincoln Cathedral, which was 160 m (525 ft) high, but when this fell down in 1549 the Pyramid ruled again. It was only surpassed in the late 19th century.

828 meters tall

160+ stories

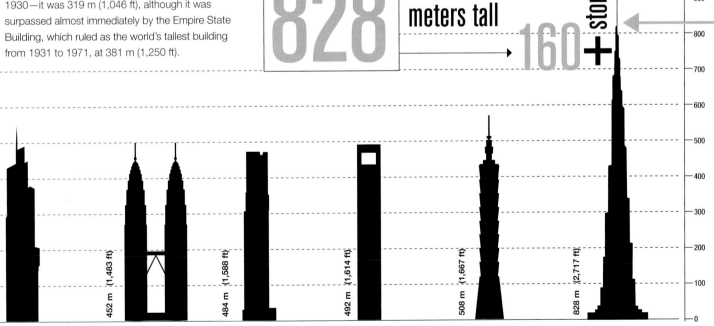

450 m (1,476 ft)	452 m (1,483 ft)	484 m (1,588 ft)	492 m (1,614 ft)	508 m (1,667 ft)	828 m (2,717 ft)
Nanjing Greenland FInancial Center, China 2010	**Petronas Towers, Malaysia** 1998	**International Commerce Center, Hong Kong** 2010	**Shanghai World Financial Center, China** 2008	**Taipei 101, Taiwan** 2004	**Burj Khalifa, Dubai** 2009

The Great Pyramid on minimum wage

▶ If the Great Pyramid were built today, using workers paid only minimum wage, it would cost over $26 billion for labor alone.

The Great Pyramid of Giza is the oldest, greatest, and only survivor of the seven wonders of the ancient world. According to the Greek historian Herodotus, who visited Egypt c.450 BCE and recorded local lore, the pyramid had taken 100,000 men 20 years to build. With a 35-hour working week, this adds up to a staggering 3.64 billion working hours. The U.S. federal minimum wage, according to the U.S. Dept of Labor, is $7.25/hr; at this rate the Great Pyramid would have cost over $26 billion dollars in labor costs alone to construct. The materials would be another matter—sourcing, cutting, transporting, shaping, lifting, and laying over 2 million blocks of limestone would be astronomically expensive. A cheaper option might be to use concrete, the material that made mega-projects like the Hoover Dam feasible. The Great Pyramid is smaller than the Hoover Dam by volume, so around 2 million cubic meters of concrete would be sufficient. At modern prices of around $130/m^3, this adds up to $260 million—relatively cheap for one of the most enduring monuments in human history.

The Hoover Dam is even **more massive** than the Great Pyramid. It consists of around 2.3 million cubic meters (over 3 million cubic yards) of concrete—enough to cover 10 square kilometers to a depth of 30 cm (or four square miles to a depth of a foot—equivalent to coating Central Park in New York with over a foot of concrete), or to build a 1 m (3 ft) square tower projecting 2,740 km (1,700 miles) into space.

The Giza necropolis outside Cairo includes three massive pyramids— The Great Pyramid of Khufu, the Pyramid of Khafre, and the Pyramid of Menkaura—and smaller tombs known as "Queen's pyramids."

Herodotus was probably misled—modern estimates suggest that a permanent workforce of less than **20,000**, swelling to 40,000 at peak times, could have constructed the Pyramid in

10–15 years

The work was probably carried out only during the Nile flood season, when most of the Egyptian population was idle anyway.

2,300,000

Supposedly 2,300,000 blocks were used to build the Pyramid—enough to build a wall around France.

More recent geophysical surveys suggest, however, that much of the interior may be filled with rubble/sand, and it is also possible that it was built around a massive natural outcrop of rock.

hectares

5¼

The Great Pyramid covers 5¼ hectares (13 acres); in the same **footprint** it would be possible to fit St Peter's Church in Rome, Westminster Abbey and St Paul's Cathedral in London, and the cathedrals of Milan and Florence.

6–46

million tonnes

The weight of the Pyramid is not known, although estimates vary from

144,000 casing stones

The **outer layer** of the Great Pyramid of Giza was covered by around

Each weighed around **15 tonnes**, was over 2.5 m (8 ft) thick, polished to an accuracy of 0.24 mm (¹⁄₁₀₀ in), and placed at perfect right angles to the stone blocks of the main pyramid structure.

7

It was never possible to see all seven of the **ancient wonders** because by the time the last one, the Colossus of Rhodes, was built, the Hanging Gardens had been destroyed (if they ever existed). The Colossus itself was the shortest-lived.

	2500 BCE	2000 BCE	1500 BCE	1000 BCE	500 BCE	0 CE	500 AD	1000 AD	1500 AD	2000 AD
Great Pyramid	■———■									
Hanging Gardens of Babylon						■—				
Temple of Artemis						■—————				
Statue of Zeus at Olympia						■—————				
Tomb of Mausolus at Halicarnassus						■————————————————			
Pharos Lighthouse of Alexandria						■—————————————			
Colossus of Rhodes						▪				

The biggest machine in the world

▶ The Large Hadron Collider is the largest refrigerator on Earth, the emptiest place in the Solar System, and the hottest place in the galaxy.

The Large Hadron Collider is a giant scientific instrument constructed by the European Organization for Nuclear Research (CERN). It is a giant circular tunnel, 100 m (328 ft) down and 27 km (16.5 miles) long, straddling the border between Switzerland and France, near Geneva. Nearly 10,000 magnets accelerate charged particles around this loop at 99.99% of the speed of light until they smash into each other with colossal force, recreating the conditions one microsecond after the Big Bang and allowing researchers to glimpse what happens when the current laws of physics break down. The LHC is the biggest machine, the biggest instrument, and the biggest experiment in the world; everything about it is superlative.

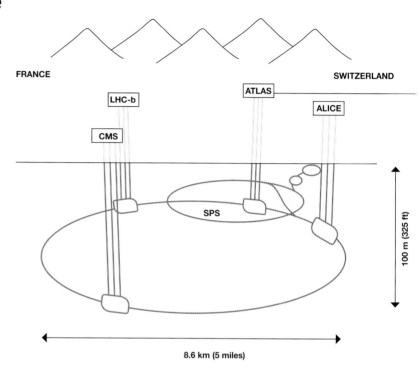

Between Lake Geneva and the Jura Mountains, the LHC is a massive ring buried 100 m below ground. Four massive detectors observe collisions between different types of particle, while a secondary ring—the Super Proton Synchroton—feeds beams of protons and lead nuclei into the LHC proper.

LHC-b   = LHC beauty experiment
CMS   = Compact Muon Solenoid experiment
ATLAS   = A Toroidal Lhc ApparatuS
ALICE   = A Large Ion Collider Experiment

The **LHC** weighs more than half the weight of the *Titanic*—at least

38,000tonnes

It contains 9,300 magnets, including the **world's largest** superconducting magnet; collectively they weigh more than the Eiffel Tower.

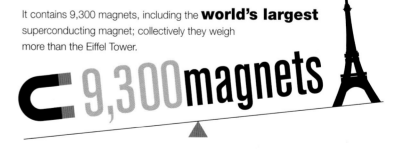

The **largest** superconducting magnet at the LHC, the Barrel Toroid, develops as much power as 10,000 cars traveling at 70 km/h (44 mph).

The magnets need to be cooled to just above **absolute zero** (colder than outer space), and the LHC's cryogenic distribution system is the world's largest fridge by a factor of

8

To ensure that the protons do not collide with any atoms inside the accelerator, the tunnel is at a **vacuum** comparable with interstellar space. The vacuum is so intense (10–13 atmospheres, ten times lower than the "air" pressure on the Moon) that it qualifies as the emptiest space in the Solar System.

When it runs at **full power**, the LHC accelerates trillions of protons around its full circumference

11,245 times a second

It produces collisions of 13 tera-electronvolts (TeV), more than

x100,000 hotter

than the heart of the Sun; for the brief instants they exist, these fireballs are the hottest place in the galaxy.

Altogether, the **collisions** that take place every second number

600 million

The amount of **data** generated by the various detectors when the LHC is in full swing is equivalent to 1% of the world's information production rate. There is enough to fill 100,000 dual-layer DVDs a year, generated at the rate of two full DVDs per second.

To handle this gargantuan flow of data, the LHC is hooked up to the most powerful supercomputer system in the world.

The LHC uses **7,600 km** (4,722 miles) of superconducting cable; if all the filaments of all the strands of the cable were unwound, there would be enough to stretch to the Sun and back five times, with a bit left over to get to the Moon and back several times.

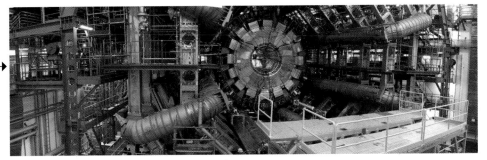

The ATLAS detector observes head-on collisions between very high energy protons. On its own, it produces enough data each year to fill a stack of CDs that would stretch to the Moon and back twice.

The ship that was bigger than the Empire State Building

▶ The largest moving manmade object on Earth covers an area equivalent to lower Manhattan, but is not as heavy as the ship that was bigger than the Empire State Building.

Mankind's ambition has traditionally outstripped its technological capabilities, but some of the machines created in the last few decades threaten to upset this arrangement. For instance, the *Seawise Giant*, a vast supertanker that went through various name changes and was best known as the *Jahre Viking*, was longer than the Empire State Building is tall, before it was broken up for scrap in Bangladesh in 2010. The *Ramform Viking* is a survey ship that trails an array of sensor streamers—collectively they cover a vast area, making this "the largest moving object on the face of the Earth," as the press release of its parent company Petroleum GeoServices proudly trumpets.

230

tonnes

Her **rudder** weighed 230 tonnes, her propeller 50 tonnes, and her steam turbine engine rated

50,000 horsepower

The *Jahre Viking* was 485 m (1,591 ft) long, weighed over **657,000 tonnes** fully laden and was too large and sat too low in the water to navigate the Panama or Suez Canals or even the English Channel. She was twice as large as the *Exxon Valdez*.

485m

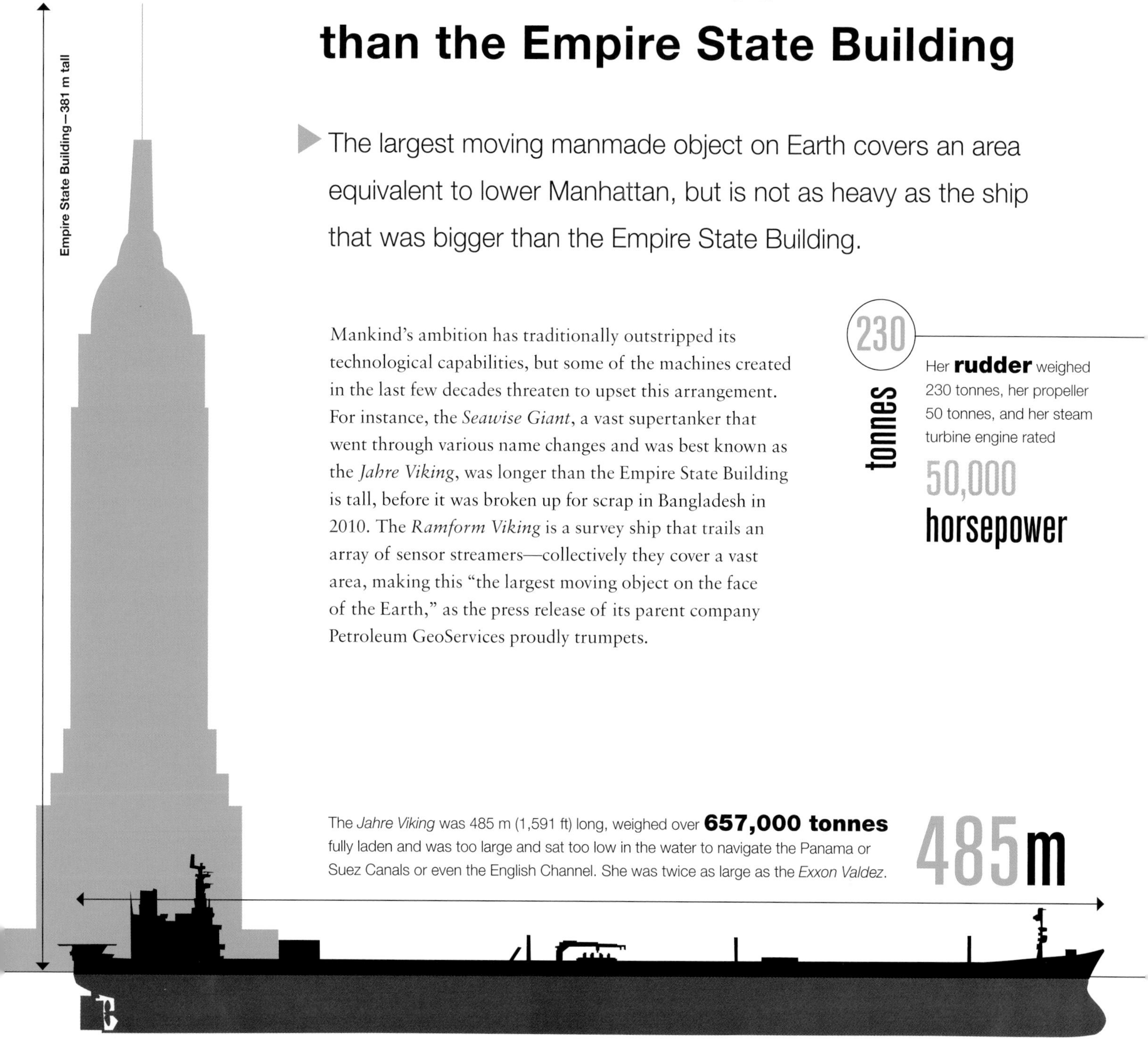

The **Ramform Viking** with its streamers deployed is more than a kilometer wide and 8 km (5 miles) long, covering 8.75 square kilometers (5.5 square miles).

The *Ramform Viking* **is a geophysical survey vessel that uses seismic waves to probe the structure of the seabed in the search for oil. The streamers carry sensitive acoustic detectors to pick up seismic echoes.**

Despite her vast size, the *Jahre Viking* had a crew of just 40—the same as two jumbo jets.

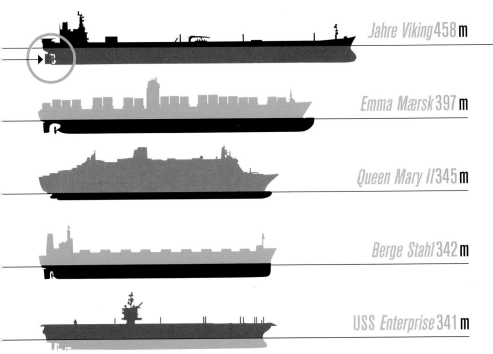

Jahre Viking **458** m

Emma Mærsk **397** m

Queen Mary II **345** m

Berge Stahl **342** m

USS Enterprise **341** m

None of these beasts is as big a machine as the **LHC** (see the previous entry), but even this would have been dwarfed by a bigger supercollider that started construction in Texas in 1991.

The Superconducting Super Collider would have been three times bigger and more powerful than the LHC, with an accelerator ring 87 km (55 miles) long and beams that could achieve 40 TeV.

After two years of construction and $2 billion of spending, however, the project was canned due to rising costs projected to spiral to $10 billion-plus, leaving nothing but a 23 km (14 mile) hole in the ground.

Another **colossal hi-tech**, extreme science megamachine is the Super-Kamiokande neutrino detector—a vast, detector-lined tank of water buried nearly a kilometer underground in Japan. The tank is 40 m (130 ft) across and 40 m high, and contains 50,000 tonnes of ultra-pure water—enough to supply drinking water to everyone in Chicago for a day.

The detector contains a billion times more atoms than there are stars in the universe.

To the Moon and back on one tank of gas

▶ If Moore's Law, stating that the number of transistors on a chip doubles every eighteen months, applied to automobile technology, a Rolls Royce would get more than 320,000 km (200,000 miles) to the gallon, be less than 1 cm long and cost less to buy than to park.

In 1965, Gordon Moore, cofounder of technology company Intel, wrote an article in which he predicted that the number of transistors that could be crammed onto a computer chip would roughly double every 24 months. Since then, this adage has become known as Moore's Law, and has been extended to cover the processing power and cost of the transistors (it has since been revised to 18 months). Despite repeated predictions that it has reached the end of its life and would no longer apply, the Law continues to hold good as chip companies press the boundaries of materials science and construct transistors on the scale of nanometers. To understand the technological and economic significance of Moore's Law, it is necessary to imagine what it would mean if applied to other technologies.

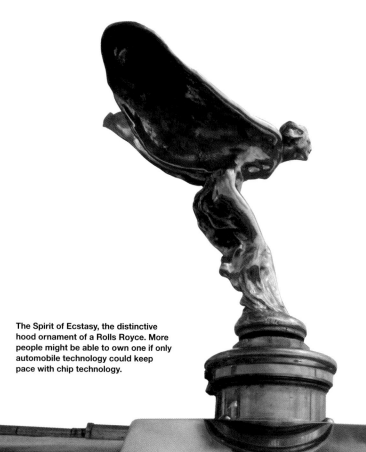

The Spirit of Ecstasy, the distinctive hood ornament of a Rolls Royce. More people might be able to own one if only automobile technology could keep pace with chip technology.

If Moore's Law applied to automobiles, a typical car that costs $20,000 today would have cost $200,000 five years ago, and would have been limited to corporate chief execs and multimillionaires. Twenty years ago a car would have cost $200 million and a car ride would have been equivalent to a rocket launch.

If automobile technology had followed the same progression that processors have achieved, you would be able to drive from San Francisco to New York (4,140 km or 2,572 miles) in about

13 seconds

In 2003, Gordon Moore estimated that the number of transistors shipped annually had reached about **10 million trillion** (10^{18})—a hundred times more than there are ants in the world.

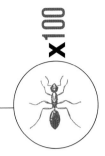

Moore's Law also means the cost of computing has fallen exponentially. A computation that "costs" $1 today would have cost $10 billion **50 years ago**.

2017 **$1**

1967 **$10,000,000,000**

Advanced chip-making tools today are precise to within one ten-thousandth the thickness of a human hair. This is equivalent to driving a car straight for 640 km (400 miles) while deviating less than 2.5 cm (1 in)—the size of the chip shown below.

In 1978, a commercial flight between New York and Paris cost around $900 and took seven hours. According to Intel, if the principles of Moore's Law had been applied to the airline industry the way they have to the semiconductor industry since 1978, that flight would now cost about a cent and take less than a second.

According to Intel, it would take you about **25,000 years** to turn a light switch on and off 1.5 trillion times, but there are now transistors that can switch on and off that many times

each second

x1,500,000,000,000

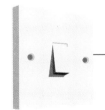

If battery technology obeyed Moore's Law, then a battery that in 1970 could hold its charge for only a single hour would today last for more than a **century**.

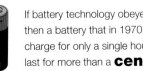

It now **costs less** to make a transistor than it does to print a single newspaper character.

Measuring
the Internet

▶ If Facebook were a country, it would be the largest on Earth.

The Internet has grown phenomenally quickly, with more than ten times as many users today as in 2000—more people are on Facebook today than were on the entire Internet in 2000. Over 2 billion people—more than the population of India or China—used the social networking site in 2017, compared to 361 million Internet users ten years earlier. Facebook gets 6 million page views per minute, 260 billion page views per month, and 37.4 trillion page views in a year.

Measuring the size of the Internet is difficult, however, partly because it is hard to know what to measure (users, websites, or bytes of data?) and partly because a key feature of the Internet is that it is distributed and not centralized, making data hard to gather.

Russell Seitz of Harvard University took a novel approach and decided to weigh the Internet. Einstein's famous equation $E=mc^2$ describes how energy has mass. Seitz applied this principle to an estimate of the total amount of Internet traffic in bytes, multiplied by the energy required to move a byte of information, arriving at the conclusion that the Internet weighs 56 g (2 oz)—about the weight of a large egg.

In the year **2000** there were just 361,000,000

Internet users
around the world

By December **2017** there were

3,770,000,000

5,000,000,000

5 billion videos are watched on **YouTube** every day

Every minute 300 hours of video is uploaded—
that's the equivalent of

1.5million

full-length films every week.

Energy has mass ($E=mc^2$), just not very much ($m=E/c^2$). Thus all the energy used by all the Internet traffic in the world is equivalent to the mass of an egg.

There are over 328 million **Twitter** users, sending

500 million tweets per day

6,000 tweets per second

205

There are 3.7 billion email users worldwide. In 2016 they sent 75 trillion emails, an average of 205 **billion emails** per day

86% **of which were spam**

In December 2009 there were **234 million websites**.

42% of Internet users live in Asia.

Google's Eric Schmidt has suggested that Google has only indexed 0.004% of the world's information so far.

Five **exabytes**, or 5 million terabytes, would be equal to all of the words ever spoken by mankind.

Google has indexed over **200 terabytes** of data so far—the equivalent of 200,000 copies of the *Encyclopedia Britannica*, or 20 times the printed collection of the Library of Congress.

The junkyard in space

▶ If the space around Earth were a landfill site, it would be full.

Spaceflight is among humanity's crowning achievements, yet it seems that we take our unenviable traits along for the ride wherever we go—including littering. Junk has accumulated from dead satellites, spent rocket stages, collisions between spacecraft, waste disposal (the Russian cosmonauts of the *Mir* space station released over a hundred bags of rubbish into near-Earth orbit), and mishaps. For instance, *Gemini 4* astronaut Edward White lost a glove during the first American space walk in 1965. It orbited Earth for a month, traveling at 28,000 km/h (17,000 mph) and becoming probably the most dangerous item of clothing ever.

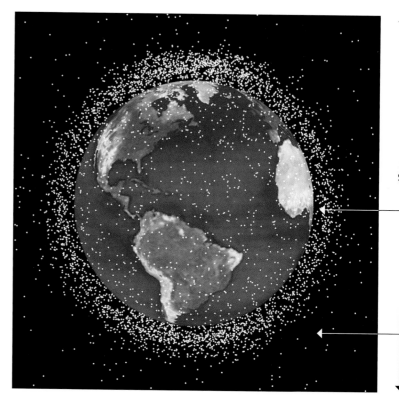

This image highlights every piece of space junk larger than 10 cm (4 in) known to orbit the Earth.

10 cm

1 cm

Particles in a **low Earth orbit** (below 2,000 km or 1,240 miles) circle the planet at around 8 km/s (5 miles/s), but once the velocity of whatever they are smacking into is added, the average impact speed of orbital debris with another space object will be approximately

10 km/s

The **fastest bullets** generally travel at less than 1 km/s. Per gram, space debris has nearly 100 times the kinetic energy—and therefore damage potential—of a tank shell.

On average, a piece of space junk large enough to be catalogued by NASA falls to Earth **each day**. So far no one has been killed by space debris reentering the atmosphere.

The higher its orbit, the longer debris will remain in space before spiraling back down to Earth. Debris below 600 km (370 miles) only lasts for a few months, but debris orbiting at more than 1,000 km (600 miles) will hurtle round Earth for a **century** or more.

The **oldest** piece of space junk still in orbit is *Vanguard I*; the U.S. satellite was launched in 1958 but only worked for six years.

Space debris rates among the most **expensive** junk ever created. Getting stuff into orbit costs around

$22,000 per kg $10,000 per lb

For the USA to launch all its terrestrial garbage into space it would cost around **$4,000 trillion a day**— roughly 100,000 times its gross daily product.

According to NASA, there are around

29,000

objects **larger** than 10 cm (4 in) across orbiting the Earth.

There are 670,000 objects between **1 and 10 cm wide** (0.4–4 in)

670 thousand

and **hundreds of millions** of particles smaller than a centimeter (0.4 in).

Getting to space is so expensive because to escape Earth's gravity it is necessary to obtain **escape velocity**—sufficient speed that, however much gravity slows you down, you are still traveling fast enough to go up.

From the surface of Earth, escape velocity (ignoring air friction) is about **11,100 m/s** or 40,200 km/h (7 miles/s or 25,000 mph).

To pitch a ball into orbit you would have to throw it **250 times** harder than the fastest pitcher that ever lived.

WWII in a single bomb

▶ The most powerful device ever created, the Tsar Bomba, exploded with ten times the combined power of all the explosives used in World War II.

Nuclear weapons are among the most terrifying but also the most impressive achievements of human technology. The most terrifying of them all was surely the Soviet AN602 hydrogen bomb, aka the Tsar Bomba, which exploded with awe-inspiring power on October 31, 1961. The most powerful nuclear weapon ever devised, the Tsar Bomba had a 50 megatonne yield; gathering the equivalent amount of conventional explosive would require a train of 666,000 15 m (50 ft)-long wagons, each carrying 75 tonnes of TNT, stretching over 10,000 km (6,200 miles).

The Tsar Bomba had a 50-megatonne yield. The eruption of Krakatoa was estimated to have a yield of

200 megatonnes

It was 1,400 times more powerful than the bombs dropped on **Hiroshima** and **Nagasaki** combined.

If it had exploded underground it would have been equivalent to an earthquake measuring 7.1 on the **Richter Scale**.

While it was exploding, its power output was equivalent to approximately

1.4%

that of **the Sun.**

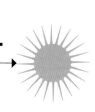

7.1

It produced a **mushroom cloud**
64 km (40 miles) high

64 kilometers

7x nearly seven times higher than **Mt Everest**.

23,360 warheads

Today, thanks to arms reduction treaties, there are an estimated 23,360 warheads in the world, with a combined megatonnage in the region of 7,000.

7,000 megatonnes

The **peak global** megatonnage at the height of the Cold War in 1973 was

27,333 megatonnes

To lay waste to the land surface of the Earth would require **1,241,166** high-yield nuclear warheads; even destroying all the world's cities would need 99,293 warheads, so we do not have the power to blow up civilization by any measure.

To overcome the gravitational binding energy of Earth and **destroy the planet** entirely would require

50,000 **trillion**megatonnes

The Chinese Room

▶ An artificial intelligence would be like a man in a sealed room, who speaks no Chinese but is armed with a big book of Chinese language rules that allows him to answer messages written in Chinese without ever understanding a word.

The Chinese Room is a thought experiment first suggested by the philosopher John Searle in 1980. He imagined a twist on another famous thought experiment relating to artificial intelligence (AI), known as the Turing test after its originator Alan Turing. The Turing test for an AI is whether someone communicating with it via written messages can tell the difference between the AI and a human interlocutor; if not, the AI should be considered intelligent in the human sense of the word.

Searle suggested a scenario where you ask questions by passing notes under a door and get answers back in the same fashion. You assume there is a person in the room reading and understanding your note and replying, but what if it were simply a machine following a complex program that analyzes your note and generates a plausible answer? Such a machine, Searle suggests, would be analogous to a non-Chinese speaker using a big book of Chinese grammar and language rules to process notes written in Chinese, generating answers in the same language without understanding a word of them. In other words, the machine might simulate intelligence but have no real understanding—what philosophers call intentionality. Critics have questioned whether intentionality is a meaningful construct, and if there is a difference between intentionality and a program complex and sophisticated enough to achieve the Chinese Room feat—i.e. any book of Chinese so comprehensive that it allowed the non-Chinese speaker to answer Chinese messages would effectively amount to the individual having learned Chinese.

Diagrammatic representation of a Turing test—an examiner sits at a terminal typing in questions.

His interlocutor could be either another person or an artificial intelligence. If he can't tell the difference, maybe there isn't one.

The coming of AI has been heralded since the 19th century and before. Critics joke that

"AI is fifty years away and always will be."

According to one estimate, based on the number of synapses (connections between neurons: ~10^{15}), and the speed at which each one processes impulses (~10 impulses per second), the human brain has a **processing power** of 10^{16} operations per second. In computer speak, this is 10 petaflops (a petaflop is a quadrillion floating-point operations per second).

10 petaflops

According to AI expert Steve Furber, it will take in excess of an **exaflop** (1,000 petaflops) of processing power to convincingly model the human brain.

Furber points out that the brain is a **million times** more power-efficient than the best processors currently in existence.

Simply **reading this sentence** out loud triggers your inner ear, independently of your brain, to carry out the equivalent of a billion floating-point operations per second (a gigaflop, or a millionth of a petaflop)—about the workload of a typical game console.

The human brain achieves this processing feat with a power consumption of around

25 watts

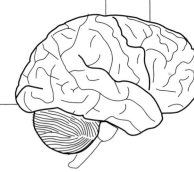

To achieve this, the games console consumes around 50 watts and generates enough heat to bake a cookie. The inner ear uses 14 millionths of a watt and could run for 15 years on one AA battery.

The most powerful supercomputer in the world is **Sunway TaihuLight** at the National Supercomputing Centre in Wuxi, China.

It has a top speed of **125 petaflops**.

It consumes **15 megawatts** of power.

Deep Blue, the supercomputer that beat world champion Garry Kasparov at chess in 1997. Deep Blue weighed 1.4 tonnes and calculated 200 million possible chess moves per second, as opposed to Kasparov's ability to plan two moves per second.

The processing power of Deep Blue is now contained within a single chip smaller than a thumbnail.

Index

Picture credits